Social Anxiety Solution

REVISED AND UPDATED

Proven Techniques and Strategies Reprogramming Your Mind to Stop Living in Fear and Stress, Overcome Panic Attack, Shyness, Low Self-Esteem, Negative Emotions and Thoughts.

Dr Henry Campbell & Dr Daniel Watson

CONTENTS

PREFACE

The pages in this book were developed through years of experiences that I have gone through, as well as what has proven to work for others that I have talked to and have researched. I also want to congratulate you for taking the time to understand your own fears and how you can overcome them.

After experiencing many different types of fears throughout my life and struggling to overcome them, I decided that I wanted to write a detailed book to help other people who are in a similar situation as I was. I also wanted to help people understand how fears form and why we experience them, because many of us have friends or relatives who struggle with fears and we have trouble understanding what is going on in their heads.

I can guarantee that you will find this book useful if you make sure to implement what you learn in the following pages. The important thing is that you IMPLEMENT what you learn. Fears are not conquered overnight but the important thing to remember is that it is definitely possible for you to overcome them. What I am giving you is the information so that you can better understand your own mind and body, as well as some steps you will need to make that journey.

Every person will experience fear in his or her life. A small amount of fear within you can be helpful in stirring up some motivation when faced with frightening situations, but hanging on to it for too long and with no good reason, can be nothing short of crippling.

As you go through these pages, you'll get a better understanding of what fear really is, where it comes from, and you'll learn several ways that you can overcome it. We will dive into what is going on in your brain, how your body reacts to your triggers, how your early childhood can influence the rest of your life, as well as what work is required of you to get past the roadblocks that you have.

INTRODUCTION

I would like to thank you for purchasing a copy of my book. Congratulations on taking the first steps toward overcoming your social anxiety and becoming a much more comfortable and confident person in your own skin. By the time you finish this book, you will be able to have a complete understanding of social anxiety disorder and the different forms that it can take, as well as the symptoms that it can cause, how you can prepare yourself and prevent these kinds of symptoms, and even how to handle them on the fly. This book will go over a number of different concepts that relate to the topic of social anxiety and social anxiety disorder, including different kinds of tools that you can use to handle its symptoms on your own, like breathing techniques and other ways to calm yourself during extreme periods of anxiety. Some of the later chapters of this book will also explain several methods of preventing these symptoms, like techniques for visualization and becoming a much more optimistic and positive person, as well as the benefits to adopting a more positive personality or outlook. These can be important things to consider with regards to handling and overcoming your social anxiety because of the nature of the disorder and how it can affect your everyday.

A significant portion of this book will be dedicated to describing the different kinds of treatment that are accessible to people who suffer from social anxiety disorder, and how you can go about seeking those treatment methods. Some of the later chapters of this book will be dedicated to different ways that you can handle your anxiety and ways that you can begin to move past and overcome this very difficult disorder on your own if you do not have access to or prefer not to see professional treatment from a licensed mental health care professional. Through reading this book you will be advised about the meanings of the symptoms that you might experience because of your social anxiety. The earlier chapters in this book, however, will be largely devoted to advising you on the basic concepts of what social anxiety is, as well as social anxiety disorder. We'll also talk about a number of different related concepts that are similar to or relevant to social anxiety, and that will be beneficial to understand for a person who suffers from this disorder.

By the time you finish reading through the chapters in this book, you should have a clear understanding of the different symptoms of social anxiety and how to handle them, whether that be by seeking out professional treatment methods, or by practicing visualization and breathing techniques to help you to overcome your anxiety. You should have all of the tools that you will need to begin to progress and grow, moving past your social anxiety and becoming a much more emotionally and mentally healthy and stable person.

CHAPTER 1:WHAT IS SOCIAL ANXIETY?

Social anxiety can often be a very confusing issue for people who do not experience it, to understand. Many are unaware that they suffer from a form of this disorder and would greatly benefit from an understanding of it. Social anxiety can be one of the most difficult to cope with and one of the most common issues in the mental health world, affecting about seven percent of the human people at any particular time. The average chance of a person developing social anxiety during the course of their life is reported at about 13 percent. However, these numbers are very likely to be inaccurate, as they only represent cases of social anxiety that have been diagnosed. Many people choose not to seek help for personal issues such as social anxiety, perceiving these kinds of problems as a sign of weakness. Some people might even believe that their issues are trivial or unimportant, as well. A lot of people will compare themselves to other people, especially those who they believe to be "stronger" than themselves, or who have gone through more difficult times. A very common phrase that might come up is "well, it could be worse," which is used as a justification for a person to ignore the problems that they face. This can be harmful, however, and it will often also make that person's problem or problems worse in the long run. Some people might also fail to understand that they have social anxiety, whether that is because they simply are not aware, or because they might be in denial about their mental health problems.

Social anxiety is a very commonly ignored and misunderstood mental illness. It is most often described as an excessive or unnatural fear of all or some particular kinds of social situations, or of any kind of interaction with other people. Most often, social anxiety will come from a sense of fear or dissatisfaction that comes from being judged negatively by other people, especially by people who they are close to and whose opinions they care a lot about. Social anxiety is a very significant problem for people who experience it, and it can cause issues like anxiety or fear that invades almost all of the different parts of an individual's personal life, even affecting things like their friendships or other kinds of interpersonal relationships, or holding them back from succeeding at work or in enjoying their hobbies. Social anxiety can sometimes also be extremely uncomfortable and difficult for people to cope with, since it is a "chronic" disease that will not be cured with time, it will take active work in order to cure.

These kinds of disorders are also usually too complex for the person who suffers from them to handle themselves. Reaching out to friends and communicating openly and honestly about the things that are bothering you can help a lot when dealing with mental disorders like social anxiety. However, professional mental health care, such as cognitive behavioral therapy, is the only method that can be advised to help you to understand and overcome mental disorders such as social anxiety. This can be thought of as similar to a medical condition; it is always best to go to a licensed medical practitioner to diagnose and treat physical conditions, just in the same way that the best method of handling mental illnesses and conditions is to have your problems diagnosed and treated by a licensed mental health care practitioner.

Social anxiety can also be a very discordant and paradoxical disease, causing a lot of conflict and turmoil within a person who suffers from it. It can cause a lot of dissonance between people who are close to someone who is suffering from social anxiety. People who suffer from social anxiety are usually perceived by others in a way that is very different from their own reality. For example, the friends or family of a socially anxious person will often perceive that person as distant in some way. They might describe that person as awkward, unfriendly, nervous, quiet, inhibited, withdrawn, or even aloof, disinterested, or hostile. In truth, however, people who suffer from social anxiety are usually prevented by their anxiety from doing the things that they want to do like enjoy their hobbies or make new friends. More often than not, a socially anxious person will want to act more friendly or communicate with other people and be more social, but their anxiety keeps them from feeling comfortable in those kinds of situations, causing them to "overthink" the situation.

Social anxiety is very commonly triggered by a large number of various situations. These can include situations such as being introduced to new or unfamiliar people or being in a space with a lot of unfamiliar people. When you are not familiar with someone, it can be difficult to understand their attitudes and intentions, and that can be daunting, making you feel judged or even on the spot. You might perceive yourself as uninteresting or worry that you make the other person uncomfortable. It can also just be difficult to "open up" and feel comfortable in a conversation with someone who is outside of your typical group of friends who you have already formed relationships with.

Being criticized, teased, or "made fun of" can also be very common triggers for social anxiety. A lot of people will be made uncomfortable by criticism or by being the "butt" of a joke, taking these things as insults or attacks on their character. In a similar way, being at the center of attention or being watched while they are doing something, especially something that they are not normally comfortable doing, can sometimes trigger feelings of stress or anxiety. When someone is watching you while you are performing a task, even one that you are comfortable with or normally enjoy, it can ramp the pressure and make you feel judged as well. This observation of a socially anxious person can also apply in situations when you have to make any kind of public statement or performance in front of a group of people. This particular kind of social anxiety is often referred to as "stage fright," and is one of the most common kinds of social anxiety. It can be seen as a common trope in film and television. Of course, this is not a complete list of triggers and causes for social anxiety or the development of social anxiety. There are a number of different situations that can trigger it, including things as simple as making eye contact with someone or being in any kind of social encounter with another person or group of people, that can cause feelings of anxiety and that can manifest in ways that will vary from person to person.

These symptoms which are either caused as a result of or which can sometimes worsen a person's social anxiety can include things like nervousness or anxiety or even excessive amounts of fear, increasing high heart rate, trembling, muscle twitches or tics, as well as other kinds of symptoms such as an abnormally dry mouth and throat, blushing, unusual or excessive production of sweat, or general feelings of discomfort or unease. These symptoms will vary from person to person in both the ways that they manifest and the levels of intensity in the symptoms that they experience.

Additionally, people who experience social anxiety will usually understand that their anxiety is irrational and that their concerns are not based on fact or reality, at least on a logical level. However, this will usually not prevent a person from experiencing thoughts and feelings of anxiety when they are placed in social situations. This can be frustrating because even if we logically understand that we are not actively being judged by other people or that our concerns are not based in reality, that does not make our anxiety go away. In most cases, we will be aware that any perceived issues are very likely to be caused by our feelings of separation or discomfort, and if we simply dismiss or ignore those feelings and try to be more confident, then the problems that we perceive in ourselves or that are created by our feelings of anxiety will simply go away. That doesn't change the circumstances or the causes of our anxiety, and will not "cure" us.

Social anxiety and social anxiety disorder are conditions which can be successfully treated with time, patience, and professional therapy. It is important to understand that these things all take time and that you will not be able to overcome your social anxiety in one day. The best way to go about this is to first begin by actively seeking assistance and treatment from a licensed mental health care professional. Social anxiety disorder and other kinds of anxiety disorders and mental health issues can be successfully treated today through cognitive behavioral therapy. It is usually recommended that you search for a specialist in anxiety disorders, who will be able to fully understand the problems that you are experiencing and will be equipped to treat them. Cognitive behavioral therapy is a very reliable and proven method of talk therapy which, if performed properly, can be incredibly helpful to people who suffer from social anxiety disorder and other similar kinds of problems. Studies have told us that talk therapy that is used for the treatment of social anxiety and social anxiety disorders can help people who suffer from these mental disorders, and they're often able to live a much happier and much more comfortable life which is no longer hindered by their anxiety or by their fear of social encounters. Cognitive behavioral therapy can be incredibly helpful in allowing people to change their thoughts, feelings, and beliefs, as well as the ways that they behave, in a very positive way. A person with social anxiety disorder who is being treated, however, must be willing and able to actively work toward bettering themselves and invest in their own growth, in order to overcome their anxiety and their fears of social situations.

Usually, this therapy will be paired with other forms of treatment like medication in order to help with the regulation of the moods and stress levels of the person who suffers from this disorder. Socially anxious people will often be given prescribed doses of drugs that have been created to ease the effects of their anxiety and stress, such as certain kinds of antidepressants. However, these drugs are usually seen as more of a temporary solution, similar to a crutch. They should not be used as the primary method of treatment, instead being used in tandem with more permanent and reliable forms of treatment, such as cognitive behavioral therapy.

Additionally, it is important to remember that every

individual person will be different, and there are not any universal methods that will always work to treat complicated issues like social anxiety. Medication can usually be fairly effective as a method of helping with the treatment of social anxiety or its symptoms, specifically for people who do not have extreme amounts of anxiety. It should also be noted that the person should not have a history of substance abuse, as these kinds of drugs and medications can be very addictive. For this reason, mental health and medical professionals will be hesitant to prescribe some of these drugs, and will usually only do so for short amounts of time in order to prevent the development of any sort of addiction or dependence on them from potentially developing. Additionally, the specific kind of medication should be important to consider. In the past, many kinds of antidepressants were promoted (usually by the drug companies that produced them) and were claimed to be very helpful in the treatment of anxiety and anxiety-related issues. This is not true, however, since antidepressants usually work by increasing the amounts of specific chemicals that are produced by the body of the person who consumes them. This can make the anxiety that a person experiences worse than they would be even without the drug, producing the opposite effect of what was intended. Medications that are more commonly and much more successfully used for and during the treatment of social anxiety and other kinds of anxiety disorders usually include depressants and other kinds of drugs that can provide a soothing effect or inhibit the production of chemicals within the body. However, not all people even need or want to use drugs to help themselves to overcome their anxiety.

One very significant change that has been made in the last decade is that the use of these kinds of medications for the use in the treatment of anxiety disorders has been decreasing gradually by the people who undergo active therapy for these disorders as well as the mental health professionals who treat those people. A large majority of these people choose to forego the use of these kinds of drugs during their treatment, focusing primarily on CBT instead.

Regardless, these kinds of medications can only help

on a temporary basis and should really only be used in extreme situations to reduce the symptoms of anxiety for the person who is experiencing them. Consuming them can be dangerous in the long run, as they will encourage the user to rely on them and focus on their symptoms as opposed to primarily treating the causes with active therapy. If drugs are used in this way for an extended period of time, they can also cause the user to form a physical dependence on them. This is very broad, blanket advice that might not be as true or applicable for every individual. You should try to talk to your psychiatrist about this topic in order to make the best decision for yourself that will allow you to overcome your anxiety as effectively as possible. You should try to find a mental health care professional who is aware of the negative effects of these drugs as well as their benefits and will be able to make an informed decision about the best kind of treatment for you and your level of comfort. You should also feel comfortable talking with your psychiatrist about the dosage and expressing whether a particular dose seems to be helping. Often, people will take the wrong dosage of prescribed drugs, which can have negative effects on your mental state and ability to overcome your mental health problems.

It is incredibly important to be aware of all of the factors that might influence your therapy and your ability to overcome your condition or conditions. However, it is also important to understand that the prognosis for these kinds of disorders is generally good. Individuals who are completing active therapy and cognitive behavioral therapy situation in overcoming their anxiety or other kinds of mental health disorders.

CHAPTER 2: SOCIAL ANXIETY DISORDERS

The official definition of social anxiety disorder has changed over the years as our understanding of it has changed and evolved, but there is a fairly consistent and reliable source of information regarding the diagnosis of mental and psychological disorders like social anxiety disorder, called the Diagnostic and Statistical Manual of Mental Disorders, sometimes abbreviated simply to "DSM". The most current edition of this manual is the fifth and is referred to as the DSM-5. The DSM and its most recent versions have begun to change the definition of many of the diagnoses over the years. The DSM used to define the diagnostic criteria for social anxiety disorder simply as a condition that occurs for an extended period of time, which was required to have been "at least 6 months". This portion of the definition of social anxiety disorder has been changed, however, and is now a little bit more specific, stating that "The fear, anxiety, or avoidance is persistent, typically lasting 6 or more months." This new definition doesn't seem to change much of the definition of this disorder for the people who experience social anxiety, but it does allow for more clarity to those who refer to the DSM-5 for guidelines on diagnostic criteria regarding mental disorders. The definition of social anxiety disorder has been made a little bit more specific in order to prevent a lot of potential misunderstandings that come as a result of ambiguous wording. The rest of the definition for the diagnostic criteria for this disorder have stayed the same.

The full DSM-5 definition for social anxiety disorder from the 2014 edition is as follows:

- The first criterion is that the individual in question will experience anxiety related to specific kinds of situations that require the person in question to interact with other people in a social context, or that will expose the individual in question to possible scrutiny or judgment from other people around them, either real or perceived. This can include things such as being watched while they perform normal tasks such as drinking or eating food, or performing in the presence of others. This will usually apply to things like singing or speaking publicly but can be applied to any kind of performance. If the individual in question is a child, their interactions with other children will usually be slightly different from their interactions with adults, so it is also important to evaluate this criterion in the context of the individual's interactions with friends and other peers.

- The second criterion required that the person in question will experience a fear of being judged or otherwise perceived negatively by the people around them if they act in specific ways that reveal that they have some form of social anxiety. This can include excessive fear of embarrassment or humiliation, as the person will often believe that these things will lead to them being rejected by their friends or families. In children, specifically, a common method of expressing this kind of fear or anxiety is through things like crying or throwing tantrums, or even some quieter reactions such as being unable to act or "freezing", hiding themselves behind an adult or someone who they are comfortable with, or being unable to speak in some cases.

- The third criterion is simply a sense of anxiety that persists during social encounters. A lot of people will avoid these situations or tolerate them for their duration, attempting to hide the feelings of anxiety that they experience. This is usually expressed in media with the common trope of hiding in the bathroom during a party or other kind of social gathering, which is not entirely untrue. A lot of people will find that quiet, private spaces like that can allow them to breathe and recover during stressful social encounters.

- The fourth criterion is that the anxious or fearful responses that a person experiences as a reaction to social situations is excessive, or disproportionate to the actual situation or threat. Sometimes, there is no real threat, and the person in question is simply perceiving a threat that they have fabricated out of their fear or is anticipating a threat in their future which causes them to become fearful or otherwise anxious.

- The fifth criterion is that the anxiety that the person in question experiences lasts for an extended period of time. Usually, this will require the condition to persist for at least six months before it can be considered social anxiety disorder. The sixth criterion also requires the anxiety that the person will feel to cause a significant amount of stress for that person. For the condition to officially be considered as social anxiety disorder, the person in question will need to be very significantly impaired by their social anxiety in a number of different areas of their life.

- The seventh and eighth criteria each refer to the sources of the symptoms of this disorder. The seventh criterion will require the person in question to be experiencing symptoms that can not be attributed to the consumption of specific substances such as recreational drugs or different kinds of medications. Their symptoms can not be caused by a physical illness if it is to be considered as social anxiety disorder, as well.

- The eighth, and final criterion requires that the sense of anxiety or fear cannot be attributed to another mental illness that has overlapping symptoms with social anxiety disorder that the person already suffers from. This is similar to the previous criterion in that it refers to the source of the symptoms that are being experienced, but is specific to mental health instead of physical health. Some of the common mental disorders are things like autism spectrum disorder or panic disorder, which both alter the ways that a person will be able to socialize and connect with the people around them, as well as things like body dysmorphic disorder. Of course, there are a large number of other kinds of disorders that share symptoms with social anxiety disorder, and which are commonly mistaken for social anxiety disorder. This is one of the most important points, as misdiagnosing these kinds of disorders can result in the wrong kind of treatment being applied, which can sometimes cause your symptoms to worsen, in some cases.

There are a few issues with this definition, as it represents a flawed understanding of social anxiety disorder which has since been changed. It can be incredibly helpful when you are trying to understand the nature of these kinds of disorders that the definitions of different kinds of mental disorders such as social anxiety have been revised over time. Social anxiety disorder, much like many other kinds of mental disorders, have only been recognized as an officially listed and valid mental disorder for a short period of dime, only having been officially recognized for the first time in the year 1980, and this disorder was also only added to the DSM in the version that was published in the year 1987, which was also the first time that social anxiety disorder would be explained adequately. The ways that we are able to understand these kinds of disorders changes over time, and social anxiety disorder is a relatively new term which we are still learning more about. Because of this, the disorder's definition also changes with our improved understanding of it, and it is able to be much more clearly and accurately defined in each new edition of the DSM. The specific criteria will also very likely be changed and altered more over time as our understanding of these kinds of disorders also changes and evolves. There are also a few specific changes to the DSM definitions that have been suggested, such as some more specific language and wording that could prevent potential misunderstandings due to ambiguous wording. Another version of the DSM, for example, references "panic attacks" that fails to clearly define the event that it refers to. It is especially important to understand the differences between anxiety, which is typically a response to a specific situation or external stimulus, and a panic attack, which is typically understood to be a warning sign, precipitating some sort of perceived oncoming medical issue or emergency of some sort, such as a heart attack, the fear of losing control in some way, or even death. Whether these fears are legitimate or warranted or not is obviously based on the specific circumstances; sometimes, the stress that one experiences during a panic attack can cause the body to experience some sort of strain, even pushing someone to a heart attack under the wrong circumstances, but they are not always omens of thee kinds of events. This can be confusing for a lot of people, and a clear definition here would be incredibly helpful.

As was mentioned briefly in the first chapter of this book, every person is different and the specific nature of the particular disorder or disorders in question and of the individual person who is suffering from these disorders can very likely affect their symptoms and can cause them to manifest very differently. There are, however, a number of more common signs and symptoms that can help you figure out if you are suffering from this disorder or even from a similar one. Some of these more common symptoms might be things like physical reactions to stress or struggle such as blushing, excessive or unusual sweating, trembling, increased heart rate, nausea or physical feelings of illness, and rigidity or "freezing up." There are also mental and emotional symptoms that can be seen, as this is a mental disorder. These symptoms include things like the mind going blank, as can be seen in cases like stage fright or if the person is put on the spot and find themselves in a situation that they are not expecting or are not familiar with. They might also tend to speak more softly or less confidently than usual, or they might avoid making eye contact with other people or avoid interacting with other people in general, especially with people who they are not familiar with or who they might be afraid or worried about being judged by. Feeling self-conscious, awkward, or embarrassed are also common, and can sometimes cause a person who is suffering from these kinds of symptoms to have a difficult time in making conversation with or interacting with other people.

Social anxiety disorder can also sometimes run in families, but the method by which it is actually passed down is still mostly unclear. In a lot of cases, some of the family members of someone with this disorder will have it while others will not. Some people might not even have any family members who share this disorder with them. Researchers who specialize in mental disorders like social anxiety have found that there are a number of different parts of the brain which are involved in emotional and instinctual responses like fear and anxiety in humans. Some researchers even believe that conditions such as social anxiety are largely caused by, or are worsened by, misunderstandings and the misinterpretation of the thoughts or intentions of other people. For example, a person who suffers from social anxiety will usually be triggered by situations when they perceive the words or actions of another person as hostile or malicious. When in reality, their intentions were simply not made clear, or they might not be particularly good at portraying their thoughts and opinions accurately. Underdeveloped social skills could be another likely contributor to issues like social anxiety. For example, if your social skills are particularly underdeveloped, you might sometimes be discouraged or feel negative about your own words and actions after talking to or interacting with other people, causing you to be apprehensive or hesitant in future interactions with those people or even with others. As our understanding of the origins of these issues evolves and becomes more clear, we can develop more reliable and effective treatments for them. Researchers also try to look for other contributors to social anxiety and other similar kinds of disorders, such as stress and other similar situations.

Social anxiety disorder is a very difficult issue to cope with. It is often the cause of a lot of unwanted and unnecessary stress in people who suffer from it, and it can even begin to cause harm to or damage the personal life of people who suffer from it or who are close to socially anxious people, like their friends, families, and even their coworkers. The DSM has a specific set of criteria for the definitions of these kinds of disorders of the mind, which are used for purposes such as forming clinical diagnoses of these disorders, and they use these criteria in order to understand whether something will fit the description of a mental disorder or if it might be caused by medical issues or other factors. The DSM specifies that the criteria for the qualification of these illnesses include things such as a pattern of behavior or psychological syndrome in a specific individual, which is not typical for that person. This means that usually, there will need to be some sort of change in the person's behaviors or mental states that are not normal for that person. These kinds of behaviors or mental states also need to be related to sources of stress that the person encounters, such as pain or trauma.

Sometimes, physical disabilities will also be considered in this context, as well. The symptoms that the person is experiencing should not be typical responses to specific events, such as the loss of a friend or family member. If the symptoms are culturally appropriate and typical for a person to experience for a period of time in response to specific events which the person in question has encountered, it can not be considered. They will also not be considered if the symptoms are revealed to simply be caused by other kinds of dysfunctions within the person, such as previously diagnosed mental disorders or physical illnesses.

These criteria apply to all mental disorders, and this includes those like social anxiety disorder. A person who suffers from this kind of mental disorder will see significant and usually harmful effects in almost all areas of their life, including all of the areas that have been mentioned above. The first area of a person's life that social anxiety will affect, of course, is their social life. They will have trouble forming relationships with other people, and they might even begin to have trouble maintaining their relationships with their current friends or family, becoming distant, awkward, or evasive.

This can cause people to isolate themselves, which can also lead to other kinds of issues like depression or a number of other illnesses, both mental and physical. They might end up suffering at work or expiring a feeling of dissatisfaction with their life. Social anxiety and other chronic mental disorders like it can permeate all aspects of that individual's lives and experiences, and if continues to be left untreated, this disorder can eventually begin to alter a person's life completely., causing them to break their connections with their loved ones or leave a job that they otherwise would have liked. The longer these problems persist, the worse they seem to get, and avoiding seeking help with these problems can be incredibly harmful, almost as much as the original illness itself.

If you decide to seek treatment, an examination will usually be performed by your doctor where they will question you about your history of mental health issues and even with medical issues before they do anything else in order to be certain that your symptoms are not caused by an undiagnosed medical health condition. If your symptoms are not caused by physical illnesses, then your doctor might refer you to a mental health care specialist, such as a counselor, psychiatrist, psychologist, or a social worker. The first step toward receiving effective treatment is to have your illness or your condition diagnosed and to work with your doctor in order to create a treatment plan that will work well for you.

Cognitive behavioral therapy (often shortened to just the acronym "CBT") is a specific kind of talk therapy that is often found to be incredibly helpful for people who suffer from social anxiety and will allow their treatment to be much more effective and helpful. CBT is intended to help a person to learn about different methods of thinking, behaving, and acting in situations that can help them to overcome their mental illnesses, and in the case of social anxiety, can help them to feel much less anxious about being in social situations. It can also help people to be able to learn and practice social skills. CBT that is conducted in a group context with multiple people can also be incredibly helpful, especially to people who struggle with social anxiety or other kinds of disorders that can affect a person's social life. A lot of people with social anxiety can also find benefits to support groups and other similar communities of people who struggle with the same kinds of disorders. In a group of people who all have social anxiety disorder or some other illness that can affect their social life, people with the illness can be a little bit more comfortable knowing that the other people in the group understand their anxiety and can relate to the struggles that they face. In this kind of therapy, it is often much easier for people struggling from mental disorders to be able to recognize that their thoughts and their fears related to their disorder are untrue and not based in reality or that they might be distorted in some way. It can also help you to understand how other people who suffer from the same or similar issues as yourself will tend to manage their issues and overcome their fears and disorders.

The most common kinds of medications that are prescribed to help with the treatment of illnesses like social anxiety disorder have also been proven to be helpful in short term uses and in smaller doses. Some professionals will use other kinds of medications, but the three that are seen most commonly are known as antidepressants, anti anxieties, and beta-blockers.

The medications that are developed specifically to treat the symptoms of anxiety, also commonly referred to as anti-anxieties, anxiolytics or antipanics, are medications that treat the symptoms of or inhibit the symptoms of anxiety. These kinds of drugs can be relatively strong and will usually start to take effect almost immediately after they are consumed to help lessen the symptoms of anxiety. Unfortunately, though, these kinds of drugs can also be very addictive if the user begins to rely on them more heavily. Many people will eventually develop a dependence on these drugs if they are taken for longer periods of time and in high dosages, which can cause that person to require increasingly higher dosages over periods of time in order to achieve the same relief from the symptoms of their anxiety. Because of this potential addictive quality in these drugs, they are typically not meant to be prescribed in high dosages or for extended periods of time, and should only be used in emergencies or as prescribed in order to avoid developing a dependence. A lot of doctors will also try to avoid prescribing these kinds of drugs for patients who have not reached adulthood yet for similar reasons. Antidepressants can be defined as anything, especially a drug, that is used to prevent or treat depression. These kinds of drugs are most common and will be the most helpful for the treatment of the symptoms of depression or of similar issues. Antidepressants work very differently from anti-anxiety medications, and because of this, are not usually very helpful for the symptoms of anxiety. Antidepressants can sometimes take several weeks before they begin to take effect, which is not helpful for immediate relief, which is usually necessary during stressful events such as panic attacks or periods of extreme anxiety. These kinds of drugs also act as stimulants, whereas anti-anxiety medications will usually be depressants, working by limiting the production of certain chemicals or by providing a calming effect to the person who consumes them, lessening the effects of that [person's anxiety and similar extreme responses to external stimuli. Antidepressants, on the other hand, will usually work by increasing the amount of certain kinds of chemicals which are produced by the body, lessening the symptoms of issues like depression. Additionally, these kinds of antidepressants can sometimes have some side effects or other problems like insomnia or trouble with sleep and altered or unstable sleeping patterns, nausea, or headaches. The side effects that people will usually experience from the use of these kinds of drugs are not usually too severe, especially if they are consumed in lower dosages. However, if the dosage amounts for these drugs are increased over a period of time, the person might end up developing a dependence on them. The third kind of drugs which are commonly used to help with the treatment of social anxiety is beta-blockers. Beta-blockers are drugs which are used to prevent or lessen the effect of many of the symptoms of anxiety by regulating the body's responses to the chemicals which cause these kinds of symptoms. Beta-blockers have been shown to be incredibly helpful for more mild kinds of anxiety such as "stage fright" or social anxiety because of this and can be very effective for these kinds of issues. They also have fewer side effects, which can be helpful for people who are trying to recover from and move past a mental disorder.

You can work with your doctor or mental health care professional in order to determine what the best medication or treatment plan might be for you, as well as the correct dosage and length of time that you should be on that medication in your treatment. A lot of individuals who suffer from anxiety and anxiety related disorders will often receive the most effective tools for their recovery and to overcome their illness through a combination of psychotherapies like CBT and medication. Each person and case will be different, however, so it is important to find the correct dosage that works best for you. You should try to communicate with your doctor and discuss any side effects that you might be experiencing as a result of the consumption of these kinds of drugs, as well. They will be able to address your concerns and potentially change the kind of medication that you receive in order to prevent some potentially negative side effects. It is important to remember that both psychotherapy and medication methods can take time to work. A healthy lifestyle can also help fight symptoms of anxiety. Things like exercise, sleep, and a healthy diet can also be much more helpful than you might think, and you should always try to reach out to your friends and family for support as well when dealing with these kinds of issues. One of the most significant issues for people with these disorders is effective communication.

CHAPTER 2 PART 2: THE SKILL OF OVERCOMING ANXIETY

To overcome low self-esteem, it is important to evaluate your logic and rationale. This way, the mistakes you make in your thinking that are driving you to make exaggerated, negative, assumptions, and conclusions. This process is called cognitive restructuring and can be used as an individual treatment, as a part of a wholesome CBT treatment plan, and as an additional treatment to exposure. Exposure is, as you learned, the ultimate way to improve your self-esteem and overcome anxiety (Antony, 2000).

Exposure is in theory, a simple concept. It requires you to face a situation that you fear and learn positive lessons from experience. This is easier said than done, as you're about to learn. It is a lesson that I learned through my efforts to support my recovering sibling. What I learned during my studies was that exposure needs to be carefully designed to be effective. If not, it can be counterproductive. As it turned out, exposure requires meticulous planning, constant self-reflection, and open-mindedness to question your beliefs. This is yet another demanding task since most of us are inclined to hold on to what we believe. We believe what we perceive to be accurate and right, and challenging that leaves us feeling confused, and often lost.

Here's a simple depiction:

If you believe that your abilities are low, you have an idea about what your life and activities should look like to live and survive with that truth. But what if you're wrong? What if you're talented and smart? What if you're even above that? Considering this idea means stepping into a headspace where you no longer understand who you are. Because your life and self-perception are made on the grounds of core beliefs, shaking those can also cause turbulence in the way you see yourself and your life. In this chapter, we'll review the process of performing and applying exposures so that they indeed teach you lessons of self-worth.

How to Plan Exposure

Exposure is also a treatment in which you purposely create a situation that tests your fears. This can be an effective way to test and overcome negative evaluations and assumptions. However, to ensure the beneficial effect of the exposure exercises, one must first create a plan. Planning exposure is necessary to identify the problem, set goals, and define evaluation standards. With that in mind, the three main parts of planning exposure treatment include (Hope, 2004):

Identify the Problem

First, you'll need to examine your thought process to identify what situations trigger anxiety, which thoughts preceded it, and what core beliefs are driving your logic. The mindset needs to be scrutinized to determine and successfully challenge fears and review the perceived risks from social events. People with SAD tend to have a perception of low emotional control, meaning that they feel powerless to manage their emotions. One of the ways to correct this is by creating a distance between the perception of your emotional state and the objective view on yourself from an observer's point of view. This helps you understand that others can't see that you are shivering and that your hands are sweating, relieving the fear that everyone notices your symptoms.

Prevent Safety Behaviors

Next, alongside your therapist, you'll review inadequate coping mechanisms that you use to cope with social anxiety. Safety behaviors are successfully treated during exposures and usually eliminated after a certain number of repetitions. You will detect and review the actions you take to relieve your anxiety, identifying those that shouldn't be used in exposures. Next, you'll work on planning to prevent or reduce, post-event rumination. Post-event rumination treatment is done by learning how to process events using rational reasoning, such as whether your recollection of the event is factually accurate, what is the evidence of the (in)accuracy, and whether or not potential mishaps have any true impact on you and your life.

Design Exposures

To design exposures, you will plan out actions, like going to places and planning events that challenge your false beliefs. This means that the situations need to be specifically designed to target core beliefs, such as:

- That you don't know how to act in a situation
- That other people will be overly critical and negative towards you
- That a social situation will have a catastrophic outcome due to your inadequacy
- These exercises are also great for you to learn that you can cope with the symptoms of anxiety and that the other people aren't noticing them as much as you think they are

For example, you might notice that you don't have to present yourself as perfectly as you thought to be accepted in a group. Another form of exercise includes engaging in a situation while making sure that no safety behaviors are being used. These behaviors make you think that you are somehow relieving the symptoms and protecting yourself from the negative outcome, but this isn't true. These behaviors are preventing you from facing the situation in the way in which you should. This way, you not only survive a situation but also feel good about yourself. These situations may include anything from spilling a drink on purpose or revealing information about yourself that you used to believe people will judge you on.

Relaxation Training

A significant portion of exposure therapy is to learn how to manage physical symptoms of social anxiety. Relaxation training helps you not only relax and feel better but also learn to control mental oversensitivity. Mental oversensitivity becomes high in social situations, and learning how to relax will help you go through the situation a lot easier. Multiple exercises can be used in relaxation training.

Some of them target relaxing specific muscles, creating an exposure ladder, and doing the so-called homework, which requires you to write down thoughts and observations on the exercise you perform without a therapist.

With muscle relaxation, you are encouraged to focus on specific muscle groups within your body, noticing the difference in how they feel when they are tight and relaxed. This form of body scanning helps you notice the muscle tension and record how it feels to be relaxed. Relaxation training can also be used in the form of cue-controlled relaxation, which means that you use the word "relax" over and over while you are in a relaxed state. Then you move on to use this approach in your everyday activities.

For example, once you notice that you feel anxious, you then take a couple of deep breaths and repeat yourself to relax. As a result, you learn how to manage your symptoms better and get a hold of yourself when you start to feel like your symptoms are getting out of control.

This technique seems simple, but it needs to be exercised frequently to be effective. It will help you learn how to manage the physical sensations when you are distressed, and eventually, learn how to relax very quickly during everyday activities. After that, you may apply this exercise to the situations that provoke anxiety. As a result, you learn to manage feelings in anxiety-provoking situations.

GIVE UP YOUR SAFE ZONE

Anxiety is constantly telling us that we have to change our external environment to soothe an internal problem. However, the opposite is true, when we heal ourselves internally, and meet our fears and traumas with authenticity, the energy that our mind has been using to shield us from those fears can now be used for actual productivity and creation of the life that you want. Change your internal world, and you will manifest the in the physical world what you truly desire.

Step One: Recognize that you are overwhelmed and that it's okay and perfectly understandable to be so.

Absolutely everyone experiences being stressed out at one point or another in your life. You are definitely not alone in your experience. Admitting that you have stress is the first priority, if you can't come to the realization that you need help, you'll never get out of the thought patterns that are disrupting your life. These thoughts can have significant damage to your mental health and getting out of the maze of thought patterns has infinite benefit.

Step Two: Think of the stress reaction as preparation for the circumstance that is arising, rather than stress and anxiety being inherently bad and uncomfortable feelings, stress is necessary and valuable to our survival, and should be appreciated as the tool it is. Ask how you can use this energy to overcome the situation.

Step Three: recognize that having to constantly think about and monitor yourself to avoid the judgments of others is exhausting and that it's time for a change of pace. The impossible task we give our minds of having to shield us from perceived threats- questioning if we said the right thing, thinking about what other people think of us, our minds are on hyper-alert because of trauma, even acute trauma will keep us feeling as though we have to prepare for a catastrophe constantly. Our perception of someone else's perception.

Step Four: Get outside of your own mind. I don't mean to push away or ignore any feelings, but to consider all the lives, perspectives and experiences that are occurring right now. We have no way of knowing all that other people are going through. Stress and anxiety can make us feel isolated, feeling like everyone around us don't understand the discomfort we are experiencing.

HOW TO FEEL SAFE WHEREVER YOU ARE

Safe attachment, which is considered necessary for a healthy state of mind, is nurtured by unconditional love, responsiveness, and support from the main caregivers.

Usually, people who have anxiety fall in one of the

three remaining categories that represent insecure attachment: Avoidant, anxious, and disorganized attachment share a different degree of disconnecting oneself from their own sensations, suppressing. This happens when, for different reasons, you start to withdraw from asking love, support, and company when you need it.

SAD links to insecure attachment styles because you form an ambivalent relationship with people: On the one hand, you want to be close to friends, family, and coworkers, because you need love, friendship, and company. You shy away from making these connections because you feel unworthy of acceptance. Feeling unworthy of acceptance goes back to early infancy, when crying for consolation or shouting out of confusion and insecurity brought you shame and criticism. However, as a conscious person, you can practice making connections to overcome emotional blockages due to feeling unworthy. You can do this in following ways:

- ***Socialize, even when you don't feel like it***. The only way for the anxiety to go away is to persist socializing despite feeling awkward.

- ***Practice seeking emotional support.*** Overly self-critical people tend to deny themselves help and support when they need it the most, further worsening the feeling of detachment from the outside world. Commit to contacting at least one person when you feel upset. Choose a friend or a family member who can understand your condition and is capable of giving rational feedback, but still cares about you and can offer understanding and consolation.
- ***Don't refuse help.*** Chances are that you rarely, if ever, ask for help at work or at home, and delay seeking medical help until your symptoms become critical. Refusing help is typical for all forms of anxiety and depression. The desire to do everything by yourself arises from feeling incompetent and from having an exaggerated perception of standards for competence and success. Learning how to ask for and accept help will help you normalize your social standards, and bring you closer with people around you.

CHAPTER 2 PART 3: HOW TO DEAL WITH A FEAR OF FAILURE

Fear is one of the basic human emotions. It is run and controlled by the part of your brain in charge of the most primitive physiological responses, called the limbic system. This system causes your mind and body to activate the "fight or flight" response, designed to save a primitive human from imminent danger. While fear is a response to a present threat, anxiety is oriented towards a future threat that can't be prevented or stopped. In other words, when you're anxious, you feel a sense of danger, and you feel like there's nothing you can do to stop it. You fear negative events coming your way that are beyond your control, and you feel defenseless.

Anxiety goes hand-in-hand with rumination, which is

the habit of obsessively thinking about unpleasant scenarios and picturing situations that can occur. This puts you in a state of constant worry, which combines with physical symptoms like pain, headaches, muscle tightness, and increased pulse.

Severe anxiety can cause a lot of emotional pain and interfere with one's quality of life and performance. However, that doesn't mean fear is always harmful. To a certain degree, it helps you act responsibly and take good care of yourself, as you're aware of risks that might come from certain behaviors. To a certain degree, anxiety helps you protect yourself from threats and meet the challenges of daily life prepared.

Anxiety and fear are all normal emotions that everyone faces. Typically, they decrease with time — however, people with SAD face chronic physiological responses that intervene in their daily life. They have the purpose of keeping you prepared for future events and are helping you protect yourself from threats. Getting rid of anxiety and fear ultimately shouldn't be your goal; you should aim to reduce the responses so that they don't interfere with your life.

HOW TO NO LONGER CARE WHAT PEOPLE THINK OF YOU

Everyone makes mistakes. However, the majority of people are able to recover and bounce back. If you have low self-esteem, it may seem like the way you feel in a particular situation determines the way you think of yourself in general. The way you feel about yourself and what you believe about yourself depends on the situation. So it's very easy for you to spiral into negativity. Those who have healthy self-esteem can evaluate themselves accurately and estimate their strengths, weaknesses, and they also believe they are worthy people, regardless of those.

Self-esteem is a personality trait that reflects your sense of self-worth. If you have high levels of self-esteem, you have a healthy sense of your worth. You understand that you deserve good things in life, such as happiness, love, satisfaction, and success. You don't expect or feel entitled to the good things in life, but you profoundly believe that you are worthy of them. People with healthy self-esteem are usually the happiest and most successful, as they tend to measure and evaluate their performance and actions accurately. However, this sense of self-worth isn't grandiose. You don't believe that you're better than other people in any way; you think that you are good enough, capable of doing both right and wrong.

People with a grandiose self or sense of self-worth are often called narcissists, while those with a deficient feeling of self-worth tend to suffer from depression and anxiety. If you have SAD that means your self worth is most likely compromised. You should be working towards improving self-esteem and self-confidence. As you can see, neither extremely high nor extremely low sense of confidence is healthy. You should work towards developing a healthy sense of self-worth that is a balanced measure of understanding your good traits and personal strengths, as well as limitations and challenges.

FEAR OF DRIVING AND FLYING

Each and every person, at some point, becomes afraid. People get scared of different things such as heights, death, driving, ghosts, being alone, and even a fear of the future. Fear is normal. It is a reaction or instinctual response when you feel like there may be a physical and emotional threat and/or danger coming your way. Fears are essential to our survival as a human race. Besides, you wouldn't be able to protect yourself from any kind of threat if you didn't feel the emotion of fear.

However, fears can become a problem when people fear circumstances that are far from any real danger and are just imaginary scenarios of the future playing out in their mind. Through years of research, we've learned that there are certain experiences, sometimes called traumas, which can happen in a person's life to generate a fear response much later in that person's life.

Pinpointing these events can range from being very obvious to extremely difficult. The good news is that the way to overcome these fear responses is within our own power. It comes down to pinpointing the events and then slowly exposing yourself to similar situations. This is often what psychologists, psychiatrists, and psychotherapists do when you visit them.

Scientists have also learned that fear of the unknown is a natural and basic human emotion that is meant to keep us safe as individuals and as a species. For example, historically, you would benefit greatly from having the fear of being lost in an unknown village. This kind of a fear would help us hundreds and thousands of years ago because societies were much less safe and had little to no connection to one another.

If you got lost back then, you might not be able to communicate with the other societies because of a language barrier or even worse, you may die of a disease because you are not immune to the diseases of this mysterious village. However, our world is much more connected today, and now that we're in a time where almost everything can be tracked or translated between cultures, this fear of being lost in an unknown village may still be there, but it has no real benefit to us anymore. In our current world, you can go into most countries and as long as you have some form of currency, you can survive.

Another example of this would be the fear of approaching a member of the opposite sex that we are interested in. This one here often holds many people back from living the kind of life they want. It is important to remember that when we lived in much smaller societies of 50-100 members, it would benefit us to have a fear of approaching an unknown member of the opposite sex because we could be risking our safety or even our chance of reproducing. In those days it was very important to be aware of these things because inbreeding and social hierarchies were key to our individual survival, as well as the survival of our tribe.

This also had a huge impact on our social status, if we were to mess up with a certain member of the opposite sex that we were interested in. Could you imagine how the other members of the opposite sex would judge you if you developed a bad reputation with another? All of your potential mates could reject you, just because you made a few mistakes or were not mature enough at the time. But in the world we live in today, you could live in New York City and not see someone you dated for months or years!!

So how does that fear of approaching the opposite sex benefit us in today's society? It doesn't. In fact, as long as you are respectful to that person and the environment you are in, you can approach whoever you are interested in with really no consequences at all! So this fear doesn't really have a use in our current overcrowded society. Unfortunately, many people haven't been able to live fulfilling dating lives because they allow their fear of approaching a member of the opposite sex to override their own desires of happiness.

Depending on your point of view, you may consider these kinds of fears as a hindrance towards your own personal growth. As a human being, you will undoubtedly begin to feel fear at times, but you have to be cautious not to allow it to negatively influence any of your actions or decisions in life. Unfounded fear can usually jeopardize your chances of success, for it can be a controlling and manipulating factor in life.

Fear can also be deemed "real" or "imaginary" and is felt due to the interpretation of a situation. It can make you feel anxious and pessimistic and incapable of thinking rationally. You'll tend to procrastinate or become doubtful and let it set some limitations in terms of your own abilities. It is also important to note, two different individuals can view the same situation completely differently.

If person A and person B have both never been in a plane before, they are both completely new to the phenomena of flying on a plane. However, if person A views the action of flying in a plane to be a new experience that will expand their view of the world and person B views this action as an extremely risky situation that he/she should avoid because of the movies they've seen, person A will feel little to no fear when the plane is starting up while person B could be terrified the entire trip. By keeping this in mind, you will be more in control of how you feel in a new or different situation.

One key is to never let your mind settle on negative thoughts that can lead to fear. Instead, fill it with courageous thoughts and your fears will eventually fade away. Courage is not really the absence of fear, but instead it is being able to prevail over the fears that are there. If you allow your mind to keep on anticipating that something horrible is bound to happen, your mind will only be clouded with daunting thoughts. The same can be said with positive thoughts.

CHAPTER 3: PHYSICAL, EMOTIONAL, AND BEHAVIORAL SYMPTOMS

Social anxiety disorder can manifest in a lot of different ways depending on the specific person who is suffering from the disorder, but there are different symptoms that can be more common in people who suffer from social anxiety. Some of these common symptoms and traits can even be things that the official definition that is used in clinical settings might not consider. These kinds of qualities can include a number of different things such as a sense of inferiority for the person, an excessive sensitivity to criticism, having some sort of difficulty or otherwise struggling with assertiveness or the lack of ability to assert one's thoughts or opinions, or even a sense of difficulty with or a lack of good judgement. Many who struggle with social anxiety will also experience some sort of fear of the judgment of other people who they meet or interact with. Because of this, social anxiety disorder might cause a person to have difficulty with pressure of any kind, and they might even go as far as to actively refuse to participate in things that they feel like might cause them to be put under some sort of social pressure.

Because of a lot of these common issues with social anxiety, this disorder, and the limitations that it can impose on the person with the disorder, can cause them to have a lot of trouble in forming relationships with other people. A person with social anxiety will have issues not only in their social life, struggling to form close bonds with people and to maintain their friendships, but also with their romantic lives because this kind of anxiety related to social situations can make it very difficult to form meaningful long term relationships with people.

In a lot of cases, social anxiety disorder and avoidant personality disorder will tend to exist comorbidly with one another, making it much more likely for a person who suffers from one of these disorders to actually have another disorder they are fighting, as well. The DSM 5 defines social anxiety disorder and avoidant personality disorder as their own distinct conditions, however they both share many of the same diagnostic criteria, meaning that they share a lot of the same traits, symptoms, and causes, and many specialists of these kinds of mental disorders will argue that the two conditions should be officially recognized as one disorder. Additionally, many mental health care professionals will also argue that avoidant personality disorder is not actually its own condition; it is simply a worsened or more extreme and progressed evolution of social anxiety disorder. They will argue that since social anxiety disorder is characterized by an inability to perform comfortably in social situations, it can lead to the avoidance of these kinds of social situations altogether, in especially extreme cases, and can progress to become avoidant personality disorder, which is characterized by the lack of the ability to relate to other people. It is unclear, however, whether these two disorders are directly connected or not.

Social anxiety disorder's symptoms are also generally split between two slightly different categories. The first of these is commonly referred to as the nongeneralized symptoms, which only present themselves in very specific social situations, or specific kinds of social situations. The second group of symptoms that accompany this disorder is called generalized symptoms, which are not necessarily triggered or worsened by specific events. Instead, these kinds of symptoms are usually present in most or all kinds of social encounters and situations. These kinds of symptoms cause a sense of dread or fear of all different kinds of social situations. Usually, people who suffer from nongeneralized social anxiety will have a specific list of the kinds of situations that make them uncomfortable or that they might have a fear of. This list might be short or long and can range from as general as stagefright or performance anxiety, to much more specific triggers, such as performing in front of certain people or types of people. A lot of people might not even be aware of the "full list" of triggers for their nongeneralized social anxiety, and might still be discovering new triggers. The symptoms of these kinds of disorders and the severity of the symptoms that you might experience can also vary depending on a number of different factors in your environment or a number of other things. Your symptoms might be especially strong and noticeable in one place and time, and might be completely gone or diminished in another, even disappearing for extended periods of time or worsening for long periods as a response to certain stressful or difficult situations that you have a lot of trouble being able to cope with. Regardless, the symptoms of these kinds of mental disorders will usually persist over the course of a person's life, and the only way to be able to move past these kinds of issues is to learn and understand how to handle the symptoms when they come up and prevent.

CHAPTER4: ISOLATION, SHYNESS, AND SOCIAL RELATIONSHIPS

This chapter will be about isolation, shyness, and the relationships that you form and keep with other people, and how they are each related to social anxiety. The first topic that will be covered is isolation. Social anxiety has a very direct and a very strong relationship with isolation , with each of the two worsening the other's effects for the person who suffers from them.

The troubling relationship that exists between the concepts of isolation and anxiety can be a very hard thing to unpack and to heal. Isolation and social anxiety seem to come hand in hand, in most cases. For someone with social anxiety, their isolation, and the habits that they form because of that anxiety regarding social situations can tend to follow them and even grow stronger as time goes on. During a social interaction, whether that social situation is something like a party where there are unfamiliar people, or something simple and seemingly insignificant like a quick trip to a grocery store, they might tend to avoid these kinds of situations because of their anxiety or because of the perceived judgement that they are afraid will come to them during that task. If they do manage to convince themselves to participate in social situations and open up to the people around them, the stress that they experience because of those kinds of situations can be extremely crippling. And after they have been removed from that situation, they might even carry that stress away with them and experience a strong, growing sense of dread regarding the actions that they have taken and how they behaved during the social encounter in question, often demonizing their own actions and perceiving them as wrong for some reason or another. These kinds of fears and concerns regarding social situations can cause the person to begin to avoid these kinds of social situations, and can also begin to impact the ways that that person is able to connect with others, even damaging the relationships that they have with their friends and families. You might even begin to view social encounters as a burden or as a "hassle," perceiving them as negative experiences and preferring to spend time in isolation instead and avoiding others instead of communicating and engaging with other people. Your avoidance of situations and social encounters, however, will not get rid of the anxiety that you might be experiencing, as social interactions are necessary for everyday life and you will continue to dread any future social situations that you might encounter.

However, while social anxiety can often lead to isolation from the outside world or from social situations, the reverse can also be true. Being isolated or experiencing feelings of isolation can often be a very significant cause of social anxiety, and can cause you to want to be alone out of a sense of insecurity. Many people will do this out of a desire to avoid feeling the anxiety and the stress that can come with social situations. If this is the case, then the anxiety that comes with social situations can be much more difficult to work with, because of a sort of disinterest in the situations at the root of your anxiety.

CHAPTER 5: DAILY PLANS FOR MANAGING YOUR SOCIAL ANXIETY

Keeping a journal that you write in daily can also help you to overcome social anxiety over time. Writing about the things that you experience in your day to day life can help you to logically work through fears that you might have about social situations.

Recording your thoughts about the different kinds of situations and experiences that you encounter can help you to gain a new perspective about those experiences, and might even allow you to understand your habits a little bit more effectively, allowing you to know when you might be starting to slip back into old, negative habits or patterns of thought, and possibly even preventing yourself from falling back on them and halting your progress. Similarly, you can try writing down simple and attainable goals for yourself that you can use to measure your growth and understand what kinds of situations make you anxious and where you need to "improve," socially speaking.

Additionally, the kinds of goals that you set for yourself and what you would like to reach can be incredibly important for your growth as a person. Whether those goals are smaller or more mundane things such as being able to overcome your social anxiety and move past your illness, or if they are a little bit larger, like long term career or relationship goals, it can be very helpful for you to write them down and have concrete goals for yourself to work towards. This can make it much easier to motivate yourself, as well as making them a little bit more real and easier to measure objectively. You will be able to set multiple smaller goals for yourself to reach within the larger target that you are aiming for which can be a helpful motivational tool as well. You will be able to reach these smaller checkpoints much more easily, and they will seem much less daunting and intimidating than the large overarching goal alone, which can begin to seem pointless or impossible if you simply keep working away at it with little reward. Another very important part of setting goals for yourself is to understand where you are along that path of growth and self-improvement in order to better know how far you still need to go. Working toward a concrete goal is often much easier than mindlessly churning away at an impossible and endless task.

CHAPTER 6: FOCUSING ON YOURSELF

This chapter will be all about focusing on yourself and your own personal growth, instead of comparing yourself to others. Additionally, when you are working toward improving yourself, and toward overcoming your mental illnesses that you might be suffering from like social anxiety, it is important to be doing that work for yourself, so that you can be more comfortable and can grow as a person. Your motivation can be incredibly important for any goal, and external goals like impressing other people can be sort of empty, and will not usually be able to serve as adequate means of motivating yourself to move forward.

You should take care to understand your motivations and the reasons that you are seeking to better yourself in this way. Maybe your daily life is difficult or stressful, and you would like to be a little bit more comfortable around the people around you, such as friends or family members.

Physical health can often be a very significant contributor to social anxiety. Improving your physical health and maintaining your body will not only allow you to feel more confident in social situations, but it can also be incredibly fulfilling! Maintaining good health can also improve your mood in a number of different ways. A healthy body can lead to a healthy mind. If your body is able to function effectively, then your mental states will also improve as your body is able to regulate itself better, and you will begin to see a lot of improvements with regards to your mental health and the illnesses that you might be suffering from. You should try to get a lot of regular exercise, mixing up the different kinds of exercise that you get, balancing a good amount of things like cardio and weight lifting or other kinds of workouts depending on your own health and the things that you might need or want to work on a little bit more. Other things can also be very helpful, like making sure that you are receiving a balanced diet, drinking enough water and staying away from alcohol or sodas. You might even want to begin to drink a little bit more tea or even small amounts of coffee. Caffeine is technically a toxic substance, and it can cause a lot of problems, such as skin irritation and the slightly more scary nerve receptor blockage that people who are against the usage of caffeinated drinks such as coffee will often cite. This is true, that coffee can cause some loss of sensitivity because the caffeine particles can become attached to your nerve receptors, after a long period of regular consumption, and you can begin to develop a dependence on it, and if you stop drinking coffee or stop consuming coffee very abruptly, then you might go through some withdrawal effects, like headaches, or your blood vessels might begin to open up more freely because the brain's adenosine receptors will not be blocked by the caffeine that you are consuming as much, which can lead to headaches. However, there are also a lot of benefits to drinking coffee, if you are able to make sure to consume it in moderation. Because coffee and the caffeine that it contains can trick the adenosine receptors in the brain, a lot of people will consume coffee if they have a headache or migraine in order to receive some slight relief. Caffeine consumption will trick your body into constricting its blood vessels, which can lessen the symptoms of headaches and migraines a little bit, as well as a number of other health benefits such as a decreased risk of things like Alzheimer's and Parkinson's diseases, higher rate of metabolism, and even a lower risk of depression. However, if you consume a lot of coffee, it is also important to remember that caffeine is mildly toxic, and that it can cause some negative effects if it is consumed in excess, so it is also very important to remember to drink a lot of water in order to flush those toxic substances out of your body in order to prevent a lot of those negative health effects.

Coffee has also been found to reduce a lot of pain after intensive workouts considerably, which can be helpful in allowing you to stay motivated to continue to exercise regularly, which can also help people who suffer from social anxiety by allowing their bodies to function as they are meant to, which can help to lessen the symptoms that you might be experiencing. If you are going to the gym more frequently, you will also be spending more time around other people without necessarily having to engage or interact with those people, which can allow you to become a little bit more comfortable with others over time! The gym is a nice neutral space to allow yourself to become comfortable with other people, because most of the other people there will also be more focused on their own workouts and won't necessarily be paying attention to you, so if you struggle with issues like social anxiety, then you can allow yourself to spend time around other people without feeling like you are obligated to socialize. Even if you are not going to the gym, you might try hiking or jogging outside, or even practicing yoga and working up that way.

Another very helpful thing that you can do is to keep regular, healthy sleeping patterns. If you are keeping a very irregular sleep schedule or staying up until the sun rises every day, that can lead to a lot of health issues, or can even cause you to feel a little off-center, which can lead to higher levels of anxiety as well.

In order to prevent this kind of anxiety, even if you are still unable to sleep at night, you should try to simply enjoy yourself in moments when you are alone. This might seem a little bit counterproductive to the goal of overcoming social anxiety, but a lot of people who suffer from social anxiety are also simply introverted people, and if they spend too much time surrounding themselves with people and throwing themselves into uncomfortable social situations without realizing their introversion, this can lead them to believe that they simply do not enjoy social situations, even if that might not necessarily be true! Sometimes, you might just need to take a little bit of "me time." You might want to spend a calm weekend by yourself every so often in order to allow yourself to recharge. You might not want to be alone, finding that you become bored or don't know what to do with your free time. If this is the case, then you can also spend your quiet weekend with one or a couple of really close friends who also understand that you just need some time to unwind. You can sit down with friends or family and watch a movie instead of going out, and this can still be an incredibly effective way to relax and unwind for an introverted person.

One of the most important things that you should remember is to simply allow yourself to be comfortable, and that you should do whatever is best for you and your own personal growth. Overcoming your social anxiety does not mean that you need to become a "social butterfly" or to always be at a party or going out. The only thing that this entails is being comfortable in social situations, at least to a reasonable degree. You can be a social person while still maintaining an introverted nature.

Additionally, you should try to remember to recognize your mental growth and reward yourself for the improvements that you have made in your recovery. Being able to move past complicated and difficult issues like social anxiety is not easy, but it can become much easier to stay motivated and continue to grow and develop, moving past your social anxiety if you are able to treat yourself every so often! You should allow yourself to be proud of your development every so often and recognize the progress that you have made by cooking a special meal for yourself, buying a new book or movie, or even by taking a little vacation if you are able to.

In that same way, it is also important to have real goals that you can work toward as well. You should be aware of what you want in your life and have concrete goals for yourself and your progress. A lot of people will actually write lists of these kinds of goals and put them up on their door or above a desk as a source of motivation or as a reminder of the kinds of steps that they need to make. You should be able to visualize what you really want and use that to continue to push forward because overcoming complicated issues like these can sometimes be difficult. You should be trying to keep your strengths and weaknesses in mind during this process as well, building on your strengths and improving on weak points where you might be lacking. You might want to start by working on the areas that will be easier to improve on, starting by working with your strengths and sticking to what you already know and eventually moving forward to the other qualities that you know you aren't as good with. Challenging yourself can be difficult (or challenging), but it will be necessary in order to overcome this kind of disorder.

CHAPTER 7:SOCIAL CONFIDENCE

A big part of social anxiety is the way that it can affect the friendships and other relationships that you form and maintain with other people. This can be difficult for people who are uncomfortable in social situations, but the key to long-lasting and meaningful relationships is effective communication. If you do not express why you are hesitant to go out every weekend or why you fail to respond to text messages every so often, then your friends might begin to think that you do not want to talk to them specifically or that you are pushing them away as opposed to simply needing to spend some time alone to help yourself to recharge. This can cause a lot of harm to your relationships and can be even worse for new friendships or ones that you are trying to form with new people. You should try to communicate effectively with other people in order to prevent your friends from misinterpreting your words and actions, as well as preventing you from being able to form new relationships. You should try to take initiative and extend invitations first sometimes, as well. You can suggest something to do to a friend, instead of relying on them to plan everything and make the first move every time. This will tell them that you want to spend time with them as well, if they might be thinking that you're avoiding them for some reason.

This can even include people who you do not necessarily consider to be friends already, who you typically view as simply acquaintances such as your neighbor or a coworker. You should try to communicate more effectively for the sake of the relationships that you already have with people you know, as well as people who you don't already know. You should try to form new bonds and relationships as often as you can. You might try to find a hobby or join some sort of cause that you believe in that can allow you to immerse yourself in social situations and meet new people. You can try taking a class of some sort or volunteering at a shelter. Getting out and doing these things can allow you to do something that you enjoy or that you believe in, as well as allowing you to meet new people who are probably interested in some of the same things that you are. This is similar to the example of going to the gym because these kinds of things can also allow you to get out and be around with people without feeling too obligated to engage or socialize with a lot of people, and you can move out of your comfort zone when you are ready to do so.

Another very important thing to understand for people who struggle from mental disorders like social anxiety is how to assert yourself. Being able to recognize and understand your boundaries and being able to say no or to express your thoughts and opinions is a very important skill and a very important part of being able to communicate well with other people. If you do not express your needs and the things that you want, then other people have no way of knowing that information and you will very likely be unhappy with the situations that you find yourself in because your interests are not being taken into account, especially if the people around you think you are already satisfied. If you think that you would prefer if things were different than how they are or if you have an opinion that simply goes unsaid, simply being a yes man can be very harmful as well. If you have a bad habit of allowing yourself to be pushed around by other people, then you should try to say "no" a little bit more often. If other people tend to make somewhat unrealistic requests of you or you simply don't feel comfortable with something, then it is completely okay for you to say no if you need to. However, you should also try to keep the opposite kind of situation in mind, as well. If you have a problem with not going outside of your comfort zone as much as you should or would like to, then you might want to start employing the "yes, and..." technique a little bit more often, or being a little bit more of a yes man. You should try to agree more often in situations that might put you a little bit outside of your comfort zone. Of course, if you are not ready to do something that is a little bit too far outside of your comfort zone then you can still say no, but this is a very important part of being able to handle your social anxiety effectively, and you should try to step outside of your comfort zone as much as you are able to.

CHAPTER 8: BOOSTING SELF ESTEEM

Being able to move outside of your comfort zone, as was mentioned in the previous chapter, is an incredibly important part of allowing yourself to move beyond the limitations that social anxiety can put on you. As you are able to do this more often and take larger and larger steps toward self-improvement and progressing toward the goal of eliminating or minimizing your social anxiety, you will be able to improve your self-esteem as well, and you will begin to find yourself feeling much more confident and comfortable as well.

One very important aspect of being able to be confident in yourself and your abilities is to have reasonable expectations and to understand yourself accurately. A lot of people will place unrealistic expectations on themselves and will strive for perfection in all aspects of their lives. This will only cause unnecessary stress and can serve to cripple yourself and hold you back from actual growth and progress. You don't need to be the absolute best at everything you do; sometimes it is perfectly okay to simply be okay. This is not to say that you should stop caring about the things that you do, or that you should simply give up at the first sign of difficulty, but it takes time to grow and develop skills, and nobody can be perfect. You should try to allow yourself to be okay with how things are, and to have realistic goals and expectations of yourself, because worrying about the things that you do will often be harmful to your growth and will halt your progress toward overcoming your social anxiety.

Surrounding yourself with positive people can also be very helpful in this way. If the people around you are understanding and patient, then you will probably be able to perform tasks without feeling pressured or uncomfortable because of a fear of being judged. You should try to understand the kind of people who you enjoy spending time with, and who will be positive and understanding of you and your boundaries. Having a strong support system can be very helpful to people who are struggling with mental health disorders like social anxiety, and might even help you to be a little bit more positive and confident in yourself if you are surrounded in other people who are the same way.

CHAPTER 9: PRACTICING GRATITUDE

Being able to be a more positive person yourself can also be a very important skill to learn, and one of the easiest ways to learn how to do this is to begin by taking note of negativity that exists in your life and replacing it with positivity instead. Positivity can be an almost "contagious" quality and can allow other people to both, directly and indirectly, gain a much more positive and light outlook on life by being around a person with a positive attitude. Adopting an optimistic personality can be difficult, but will help you a lot on your path to overcoming this incredibly difficult and complicated disorder.

One very easy way to begin to replace one very big source of negativity in your life with positivity instead is to simply stop apologizing for the things that you do. Everyone makes mistakes, and if you are always focused on what you are doing wrong, then your self-esteem will begin to plummet. Constantly apologizing like this can seem almost like a habit as well, and will almost begin to lose its meaning after a while. Additionally, most apologies are primarily about the person who is apologizing; you are saying that YOU are sorry or that YOU regret what you have done, and this essentially does nothing for the person who you are apologizing to. If you must, you can try to help the situation by fixing it or dealing with the consequences of what has happened. If you feel that you absolutely need to say something about the situation, then you should try to thank the other person who you would be apologizing to. This puts the emphasis on them, instead of on yourself, which can be much more effective, and is also much more genuine and meaningful. For example, if you accidentally bump into someone walking down a narrow hallway, then your first instinct might be to say "oh, I'm sorry." And to try to let them pass, leading to an awkward interaction wherein both people will try to let the other pass first. If you instead react to the other person letting you pass and express your gratitude toward that person for allowing you to go first, then this will allow them to understand that their politeness has been noticed and appreciated. This will allow them to feel a little bit better about themselves instead of simply having an awkward and uncomfortable experience with a stranger in the hallway. If you feel like you have inconvenienced someone or that they have gone out of their way to help you, then you should not apologize; instead, you should thank them for the kindness that they have given to you.

You should also try to stop complaining about minor inconveniences in your day to day life, and blaming other people for unpleasant or uncomfortable things that happen, especially events that occurred in the past that have caused you some unpleasantness. If you dwell on the past in this way, then you will not be able to move forward in the present. Additionally, you will not be able to take ownership of those events and take the agency that will allow you to move past them completely.

CHAPTER 10: COGNITIVE BEHAVIORAL THERAPY

One of the most common methods of treatment that is used with patients who suffer from a social anxiety disorder is a form of talk therapy which is called cognitive behavioral therapy (or CBT). CBT will usually have you working with a licensed mental health care professional in order to allow yourself to better understand the way that you think and perceive different things and be able to handle different or unfamiliar situations a little bit more effectively. Most of the time, cognitive behavioral therapy will be structured in a very specific kind of way with a more limited number of sessions which have been decided beforehand.

This kind of psychotherapy can be a very effective tool to allow you to develop better methods of handling yourself in uncomfortable or unfamiliar social situations. Sometimes, CBT will sometimes be performed on its own, and sometimes it will be used alongside other kinds of therapy, depending on the person and the kind of care that they and their illnesses might require. It can also be very effective for the treatment of complicated mental disorders like depression, anxiety, eating disorders, personality disorders, and even things like post traumatic stress. Some people who use CBT to help themselves to be able to grow and improve their mental health might not eve have a specific mental health disorder; they might be relatively mentally healthy people who simply wish to improve their general mental health and be able to handle unfamiliar or stressful kinds of situations a little bit more effectively.

The different kinds of things that people will use cognitive behavioral therapy for are to treat or manage their existing mental illness or illnesses and the symptoms of those illnesses (especially if they do not have access to- or prefer to avoid the use of- medications to help them with their mental illnesses), to be able to understand and manage their emotional and mental states more effectively, and to cope with their mental illnesses more effectively as well as the kinds of events that cause those illnesses to develop. Some of the common mental health disorders that people commonly use CBT to help to treat are depression, eating disorders, substance abuse, obsessive-compulsive disorder, bipolar disorder, and schizophrenia, as well as things such as anxiety disorders like social anxiety.

In a lot of cases, this kind of treatment is the most helpful when it is used along with other kinds of treatment methods like the use of medications or other therapy. Luckily, this kind of talk therapy has very few risks outside of a possible feeling of discomfort or unease when discussing particularly stressful events or topics. CBT will try to encourage you to explore a lot of stressful or painful kinds of emotions or memories, in some cases. Some people might even feel physically drained after leaving a particularly difficult session wherein they had to confront a lot of uncomfortable topics.

If you want to get started with cognitive behavioral therapy, then it can be an incredibly effective method of treatment. You should first start by finding a good therapist. If you are not sure where to look, then you might want to try to ask your doctor or find out if there re any doctors or particular kinds of services that your health care plan covers. A lot of people will have access to a number of different kinds of counseling services or suggestions for effective services through their employer, through an employee assistance program of some sort. If you do not have access to these kinds of services or if you might prefer to do your own research, then you can go that route too. You can look into different kinds of treatment and different doctors online or through a local psychological association in order to find out about the kind of doctors that you feel might be able to help you the most effectively.

Part of determining the specific kinds of services and the specific doctors that you will want is to consider the costs as well. Some health care plans might only provide coverage for a certain number of sessions within a specific period of time, or up to a certain cost amount per year.

Once you have found a doctor and scheduled an appointment, you should prepare for that first appointment with this new doctor. Before you go in, you should try to think about what specific topics or problems that you would like to discuss, and if there are any concerns that you might have about the kind of help that you would be receiving from this doctor. This can include things like the qualifications of the doctor that you are seeing or if they specialize in any particular area, among a number of other topics. A lot of people will usually advise people new to this kind of therapy to write a list of topics that they want to discuss with their doctor, on the first session as well as future sessions, because it can be very easy to forget these things once you enter the room. The most important thing to consider is that you are able to find a therapist who is able to meet your needs and provide the kind of care that you will require.

CHAPTER 11: PANIC ATTACKS

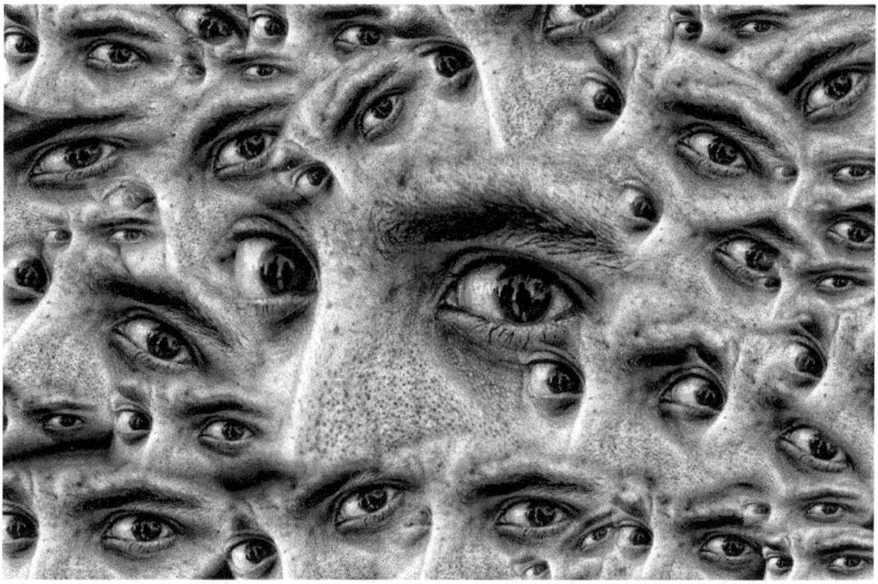

Another symptom of high levels of anxiety that you might experience are panic attacks. On a yearly basis, it is estimated that somewhere between one-tenth to one-third of the entirety of the human population will suffer from at least one panic attack. Panic attacks can be defined as very sudden or unexpected fits of stress that usually take the form of an unusually high level of fear or anxiety. These experiences are usually unwarranted or disproportionate to the external stimulus that led to it, often being much more than the actual situation calls for. Most people who experience these kinds of panic attacks will experience an unexplained or unreasonable amount of stress or anxiety, usually accompanied by things like trembling, sweating, trouble to breathe, and increased heart rate, as well as a number of other similar symptoms. Many people will even feel things like a tightness or pain in their chest, or a feeling of disorientation or detachment from the world around them, which is referred to ad dissociation. This is the "out of body" experience that people will sometimes talk about during stressful or traumatic events. Some people have even claimed to feel sensations that have led them to believe that they might have been experiencing things like heart attacks or strokes, in extreme cases.

The terms anxiety and panic are commonly used as if they were synonymous, but the two terms actually describe distinct and different kinds of experiences. Usually anxiety, especially more general forms of anxiety that are triggered by many different kinds of events as opposed to non-generalized forms of anxiety which are triggered by much more specific events, will usually last for a little bit longer. On the other hand, fits of panic will usually be much shorter, but they will also usually be much more intense. Most panic attacks will only last for approximately ten minutes, since the human body is not particularly well equipped to handle these kinds of intense and heightened states of arousal or stimulation for extended periods of time. Panic attacks will usually trigger the "fight or flight" response, which is normally a survival instinct which is triggered as a response to dangerous or hazardous events or experiences. However, a panic attack can be considered to be harmful in this way, since they do not always require a real threat to have occurred in the moments when they occur, because many of them are sparked by flashbacks to previous events, whereas anxiety will typically be a response to things that are happening around you. A lot of different kinds of things can trigger a panic attack, such as the act of exercising too intensely or too much, drinking too much coffee or consuming an excessive amount of caffeine, or even simply watching a scary movie, in some cases. However, anxiety is a much more future- or present-oriented emotional response which usually comes as a result of the anticipation of a negative occurrence in the future or the experience of one in the present. Feelings of anxiousness will usually last a little bit longer and will cause a little bit less stress on your body when compared to something like a panic attack.

Only about 3 percent of the people who might be prone to experiencing panic attacks will actually be diagnosed with the panic disorder because of its similarities with other kinds of mental disorders, and especially with anxiety disorders. Panic disorder is known for causing repeated anxiety or panic responses such as panic attacks. A person who suffered from the mental illness, which is referred to as panic disorder will be likely worried about any potential panic attacks that they might experience in the future. A lot of people will consider panic disorder to be characterized by a fear of panic attacks that the person who suffers from the disorder might anticipate. A lot of the time, this anticipation of future panic attacks will often cause a lot of fear and stress about the possibility of experiencing another attack at a particularly inopportune time, which can severely impact that person's life, often holding them back and preventing them from reaching their full potential, or even pulling them down from it in some cases. Even just the anxiety about possible panic attacks that might come in the future can sometimes cause a lot of stress on both the mind and the body, and can often lead to a panic attack a lot of the time. Usually, these kinds of experiences will be characterized by symptoms that are common with strong fear, such as an increase in your heart rate, trembling or general nervousness that can sometimes grow to become a panic attack.

When people have a high level of sensitivity to anxiety and the symptoms of it, they might sometimes interpret normal physical feelings as if they were a little bit more serious than they are in reality. These kinds of misunderstandings and misinterpretations can sometimes cause increased levels of stress as well, which can also lead to panic attacks. If you suddenly begin to feel a sharp increase in your heart rate, this can be scary, especially if you are not sure why this is happening. In these cases, you might begin to worry that you are about to or having a heart attack, which can sometimes lead to panic or general anxiety. Similarly, the feeling that a person is beginning to lose control of their life can also lead to panic attacks in some cases. These symptoms and their causes will vary from person to person, but most people will usually find these kinds of situations very uncomfortable and will usually become anxious or begin to panic.

A lot of people will believe that these issues are inherited from the parents of the person who suffers from then. While this is true in a sense, these kinds of mental health issues are very rarely inherited genetically. Issues like these kinds of mental disorders are usually developed as learned traits as a result of past experiences that a person has had, a lot of which happen with or because of family members during adolescence. Sometimes, a person might develop a fear of a particular kind of internal response, like the symptoms of panic attacks or anxiety, which is often referred to as "interceptive conditioning", and can cause some people to "teach" their bodies to force them into an excessive fear or panic state as a response to normal changed in their physical states that can occur throughout the day like increased or decreased heart rate or other kinds of things. This is the reason for a lot of habits that people might develop which are related to their mental illnesses. Some people can experience nocturnal panic attacks for this reason, even without having a nightmare. It appears as if there's no trigger, but the trigger is these slight body changes. Luckily, this kind of sensitivity and the conditioning that can cause anxiety and panic in people who have developed these kinds of habits can be treated and handled very effectively with specific kinds of talk therapy like cognitive behavioral therapy.

There are a number of different kinds of things that can lead to a person being a little bit more likely to develop these kinds of habits through interceptive conditioning. If you are able to understand these factors, then you will probably be a little bit more likely to help yourself or other people around you to avoid developing habits like these, which an be very helpful for people who suffer from anxiety, and especially social anxiety and social anxiety disorder. Traits that are inherited biologically from your parents or families can make a person a little bit more likely, or even a little bit less likely to develop specific kinds of personality traits and styles that might lead to these kinds of habits or the development of anxiety disorders later on in their lives.

In a similar way, you can also develop these kinds of traits as a result of learned behaviors that you might have picked up throughout your life as well. This is especially true for traumatic events or experiences, or ones that happened during the adolescent or developmental stages of your life. These experiences can lead to an increased likelihood that a person will end up developing negative habits like the ones that have been discussed in this book or to develop harmful mental disorders. People who grew up in isolated environments and who were not exposed to situations that allowed them to explore the world and become familiar with the kinds of interactions that they might need throughout their lives will probably be a little bit more likely to react poorly to these unfamiliar situations and develop mental disorders like social anxiety disorder as a result of stressful or traumatic experiences. This leads to the third kind of triggered event for the development of these kinds of psychological disorders. A particularly negative experience with specific symptoms or manifestations of mental illness might cause an irrational fear of those symptoms or of that illness. This might take the form of some sort of traumatic event that the person in question experienced themselves or even as a negative event that they might have experienced witnessing a friend or a family member in their own struggles with that disorder. Usually, these kinds of things will cause an initial panic attack that will typically catch the person off guard. Not understanding how to handle their panic attack, they will usually have a particularly stressful time which will likely cause them to develop an intense fear of panic attacks that they might experience in the future. These initial panic attacks can be especially frightening if the person happens to be a difficult activity during this time, or one that requires concentration and can be made very unpleasant by the sudden appearance of a panic attack, such as driving or performing in front of other people in a social situation.

A lot of people who suffer from these kinds of mental disorders will also very likely only realize that they suffer from them after a panic attack. Usually their first panic attack- catches them off guard and they end up going to the emergency room because they do not understand what is happening to them and what might be causing their symptoms. One popular reference in recent media is from a movie, but can be a very apt comparison for the purposes of understanding this kind of experience is tony stark in the third iron man movie, rushing out of a restaurant during a panic attack, thinking that he had been poisoned. Panic attacks are very intense experiences, and thinking that you might have been poisoned is absolutely a legitimate response to experiencing one for the first time. However, most emergency rooms are not very well equipped to help to ease the symptoms of anxiety and panic disorders. Most medical facilities should have some drugs that can provide immediate, short term relief, but they will not be able to provide counseling or other kinds of mental health care. If you are worried that you might be suffering from a mental illness or are experiencing some significant psychological symptoms, it is recommended that you consult with your primary care provider and ask them what the best course of treatment may be.

Luckily, a lot of anxiety disorders are very easily treatable and can be "cured" very reliably with a number of different techniques. One of the most successful and effective methods of treatment for these disorders is referred to as CBT, or cognitive behavioral therapy. There are other certain kinds of treatment as well, like acceptance and commitment therapy, or ACT, which have been developed as effective methods of treating specific kinds of mental disorders, such as panic disorder and social anxiety disorder.

A lot of doctor will aso prescribe certain kinds of drugs to aid in the treatment of these kinds of conditions as well. Some of the most effective and most common of these drugs are called SSRI's, or

Selective Serotonin Reuptake Inhibitors. These SSRI's are types of medications which are commonly used to prevent the symptoms of anxiety, panic attacks, and other kinds of emotional and mental symptoms caused by mental disorders by regulating the amounts of certain chemicals which are produced by the body. These drugs can be somewhat helpful, but CBT (as well as other kinds of talk therapy) is usually a lot more effective in treating these kinds of issues than simply taking medication.

While panic attacks will be most common in people who suffer from anxiety disorders like panic disorder or social anxiety disorder, they can also happen in people who do not, and can happen for people who don't necessarily meet the criteria for these kinds of illnesses, especially in trauma survivors and people with other similar kinds of histories. There is also a very common misconception about the ways that a lot of people can or should try to go about dealing with them. The most common advice people will give to others who are experiencing a panic attack is to take deep breaths, think positively, and try to calm down. However, trying to control your anxiety in these ways can backfire or even be completely counterproductive. Struggling to prevent a panic attack can have a very negative effect and will actually very likely trigger or worsen a panic attack. During a panic attack, a person should try instead to do something that might seem a little bit counterproductive; let go of control and allow the panic attack to run its course. It can also be useful to remind yourself that you can't die from a panic attack, and that it's your body's normal, natural reaction, which is not dangerous.

If it is possible, you should try not to remove yourself from the situation that you are in when your panic attack begins until the anxiety has subsided, instead choosing to confront your fears. If you absolutely need to, you can remove yourself if the symptoms are too difficult to deal with, and let the symptoms subside on their own over time as you relax and calm yourself a little bit. Regardless of how you handle it, you should try to pay attention to the situation that triggered your panic attack so you will be able to better understand how to handle it in the future. This is similar to the reason why you should try not to completely remove yourself from the triggering situation if you can handle the symptoms on your own; If you don't run away from the triggering event, you will be able to immerse yourself in it more, and you will eventually be able to familiarize yourself with that situation and become more comfortable over time. There are some ways that can help you to handle panic attacks and lessen the symptoms or effects of the attack until it passes, however.

One of these methods is to try to practice breathing a little bit more mindfully. If you are having trouble breathing, this can be especially helpful. This can be very helpful during panic attacks, since one of the most common symptoms is a tightening or pain in the chest which is caused by constricting muscles, and can tighten on the lungs or airways, causing that "wheezing" effect that some people will experience. When this happens, you should try to practice a special kind of breathing based on the diaphragm. This will essentially mean that you should be breathing with your belly instead of your chest. This allows you to breathe a little bit more deeply, as well as forcing you to focus on your breathing and take a little bit longer to inhale and exhale, which can help to slow your heart rate and allow you to calm your nerves a little bit more easily. Visualizing a balloon in your belly can be very helpful in helping you to breathe in this way as well as serving as an effective distraction in some cases, from the triggering event that caused your panic attack. You should also try to exhale a little bit more slowly than you will be inhaling, as well, in order to allow the air to flow more smoothly. A lot of people will spend about three seconds on the inhale and four seconds on the exhale.

Another kind of strategy that you can use to handle your panic attacks is commonly referred to as "progressive muscle relaxation." Progressive muscle relaxation is a technique for relaxing groups of your muscles individually. In order to do this, you will need to try to do a short examination of your body and the muscles on your body, tightening and releasing each of the different groups of muscles one by one. You can try to clench your fists hard, tensing your muscles a little bit, and let go slowly, and doing the same for your forearms, and then the shoulders and back, moving through the different groups one by one. This is very similar to when you might get a leg cramp and have to tighten your leg in order to move it to a better position and loosen the muscles, allowing the cramp to disappear.

You should also try to remember that the panic attack is temporary and that you will eventually be fine, once the sense of fear that you are experiencing subsides. At the moment, you might forget that you are experiencing a panic attack, or you might be aware but still unable to keep yourself from experiencing it. Either way, it will be very important for you to remind yourself that it is a panic attack and that this sensation and the symptoms, and the feeling of dread that you are experiencing will go away soon. You will only need to focus on your breathing and ride it out.

You might also try to use some kind of grounding techniques like coloring in a book or looking at your surroundings. A lot of people will gain a sense of comfort from coloring, as on a page in a coloring book. It can be very easy to become immersed in simple activities like this, which will make it a little bit easier for you to distract yourself, causing your panic attack to disappear. A lot of other people will say that listing the things in your surroundings can be effective. This can include things like saying the names of the things in the room with you out loud or noticing all of the individual sounds that you can hear can be an effective technique for grounding yourself during a panic attack.

CHAPTER 12: FACING YOUR FEARS

A topic that has been brushed upon in the previous chapters in this book is being able to "face your fears", immersing yourself slowly in the situations that might trigger your social anxiety or other mental disorder in order to overcome that condition more effectively than if you were to simply remove yourself from the situation and avoid it altogether.

This is a concept of "exposure therapy" which works in much the same way as this technique; You place yourself in situations or in proximity to objects that trigger an anxiety response from you in order to become better acquainted with it and lessen the amount of discomfort that you experience by being around that thing more often. For a person who suffers from an issue like social anxiety, this usually means social situations and other kinds of gatherings or interactions with other people. Ordinarily, these kinds of events would be overall pleasant experiences, but to a person who suffers from social anxiety, they will be much more negative and stressful kind of experiences. Entering into social situations like these can trigger feelings of discomfort or unpleasantness, and can lead to many of the symptoms that accompany anxiety, and can create a much more negative experience. This can cause you to begin to avoid social situations and other kinds of events wherein you might be around other people in large or small numbers. This can be very harmful to your development, and can negatively impact a number of different areas of your life.

However, there is a very effective way to allow yourself to become more comfortable in these kinds of more stressful situations. Exposing yourself to the kinds of things that make you uncomfortable, if you intend to try to move past that fear, can allow you to acclimate yourself to those things much more easily as well. This is a very large part of treating social anxiety and will be important to get used to doing. Stepping outside of your comfort zone is helpful for people who might not necessarily even be suffering from social anxiety disorder or any other mental disorder. You should always try to experience new things as often as you can.

CHAPTER 13: HOW TO STOP WORRYING AND SLEEP BETTER

Social anxiety and social anxiety disorder can also cause a lot of damage to your sleep schedule, as can most of the different kinds of anxiety disorders. You might find yourself lying awake at night unable to sleep because you are lost in thought, and unable to clear your head and relax. The National Sleep Foundation most often refers to this kind of late-night restlessness as "maintenance insomnia," and characterizes it as an inability to fall asleep, or stay asleep, as well as the inability to go back to sleep once you have woken up. This is usually caused by a chemical imbalance or by a sense of "confusion" within the body, which is why a lot of people will take things like melatonin capsules in order to help them to maintain a more regular sleep schedule. These capsules will allow the body to regulate the rates and the times at which it will produce serotonin and melatonin, which can help you to keep a much more consistent sleep schedule and get much more calm periods of rest.

However, these do not treat the underlying cause of your late night restlessness or maintenance insomnia. This can be a very dangerous pattern to develop and can cause a lot of damage to your emotional and mental states as well as simply causing you to feel poorly in general. You might end up sleeping poorly or uncomfortably, which can lead to much higher levels of anxiety as well.

There a number of very helpful things that you can do to help yourself to get o sleep much sooner and prevent yourself from having to stay up all night trying to get to sleep, stressing out about the sun coming up in so many minutes. The first thing you should try to do is to limit the amount of usage of your computers, televisions, and other electronic devices for about two hours before you plan on going to sleep. This is also true for phones. You should try not to have your phone next to you while you are trying to sleep if you are tempted to pick it up and begin scrolling on facebook or Instagram when you are bored. These kinds of devices emit a kind of light that generally exists in the lower end of the color spectrum, closer to the blues that you will see in the sky during the day time. This can trick your body into thinking that it should be awake, which is a very significant cause of a lot of those sleepless nights. You should also try to keep your bedroom light off, or only use red light bulbs during the night. The light that is emitted through those red bulbs will allow you to still be able to see while preventing your body from thinking that it is the middle of the day and that it should be awake and active, which will allow you to be able to sleep much better than otherwise. A lot of people will also wear sleep masks in order to completely cover their eyes as well, but eliminating the use of electronic devices for a few hours before you plan to go to sleep is probably one of the most important parts of a good night's rest, especially with the accessibility of our mobile phones and laptops. It might be tempting to kill your idle time with a few episodes of your favorite TV show, but this is actually hurting your sleep patterns more than anything else. Sometimes, even the noise that these shows produce can disrupt your sleeping patterns, so it is usually best to simply eliminate these external stimuli altogether.

Another good strategy is to try to eliminate any potential cause for concern before you go to bed. If you are cutting out your phones, computers, or TVs a couple of hours before you go to sleep, then that two hour period before you actually lay down can be a great time to get rid of all of the issues that you might think about when you are trying to sleep. The key with these strategies is to keep it light and avoid sources of heavy stimulation that will make you think that it is midday at that time. While you are preparing to go to bed, you might try writing down your experiences in a journal. This can help to pass the few hours before you actually lay down, and it can also be incredibly useful for people who suffer from social anxiety to write down their experiences and the things that they might be worried about. This kind of practice can help you to understand the situations that you are stressed out about and to be able to stop worrying about them a little bit easier. You can get all of the worries out while you are writing, which can be helpful for calming your nerves and assuaging any potential sources of fear or anxiety, as well as simply being a very nice and calm activity that can even be somewhat therapeutic at times. This can help you to unwind and allow you to clear your head BEFORE you lay down so that you don't have to have any of those moments when you realize that you have to take care of one last thing before you forget.

If you do these things, cutting out the use of electronic devices that emit blue light (or really any light if possible) and getting out all of your concerns by listing the events of the day to yourself or writing in a daily journal that you keep, they will almost certainly allow you to be able to get to sleep much more quickly. There is another method that you can use as well, which isn't its own method so much as just something that you should be doing; you should be allowing yourself to unwind and relax before you begin to go to sleep. You might take a shower or try to draw, or do whatever can help you to unwind. If you are anxious when you get into bed every night, then your brain will form that connection and will make you much more anxious in bed. Stop stressing over getting to sleep. Of course, this is not as easy to do as it is to say, but nothing is worse than realizing that you are struggling too much and that the fact that you are trying to go to sleep is preventing you from actually falling asleep. This can create a sort of feedback loop, wherein you just become more and more irritated and then when you finally do get to sleep, you won't be properly rested anyway because you are still anxious and tense. A lot of the grounding techniques that people will use for dealing with panic attacks and other forms of anxiety will usually work very well to help you to distract yourself and relax yourself enough to be able to sleep in most cases. In addition to this technique, cognitive behavior therapy can be very helpful in dealing with worry, anxiety, and associated sleep difficulties.

Some other different kinds of techniques that can also be very helpful are meditation, as well as other kinds of relaxation techniques. These kinds of practices can allow a person to be able to calm themselves and reduce the symptoms of their anxiety or their panic attacks. Of course, they will usually be most effective if you are already familiar with these techniques and can remember how to use them during these episodes of panic and anxiety. If you are trying to use these techniques for the first time during a panic attack if you do not already understand how they work and how to perform them effectively can be very difficult, and will very likely not work at all. Trying to do this will probably only cause you to be more stressed than before, and might potentially worsen your symptoms. If you attempt these techniques in this way, they might even end up convincing you that the technique is not helpful, even if it might be once you learn how to use it effectively.

Another very important thing to remember is to simply get enough sleep. Starting with the basics can be helpful in any situation. One should never underestimate the powers of rest, water, and exercise. A lot of people will be tempted to stay up later into the night and to sleep in later during the day, but this can be harmful to your body, causing it to be less effective in the regulation of certain chemicals, which can worsen a lot of the symptoms of anxiety and will make it a little bit more difficult to manage your disorder. Additionally, keeping an irregular sleep schedule can also make it more difficult for the body to manage itself and recover from stress, like wounds, and you will probably notice that cuts take longer to heal or that you feel very strong hangovers after nights of drinking. Being able to keep a consistent sleep schedule is very important for any person, but especially to people with mental illnesses like social anxiety disorder that can be made worse by chemical imbalances in the body.

CHAPTER 14: STRATEGIES FOR CREATING A POSITIVE MENTAL ATTITUDE

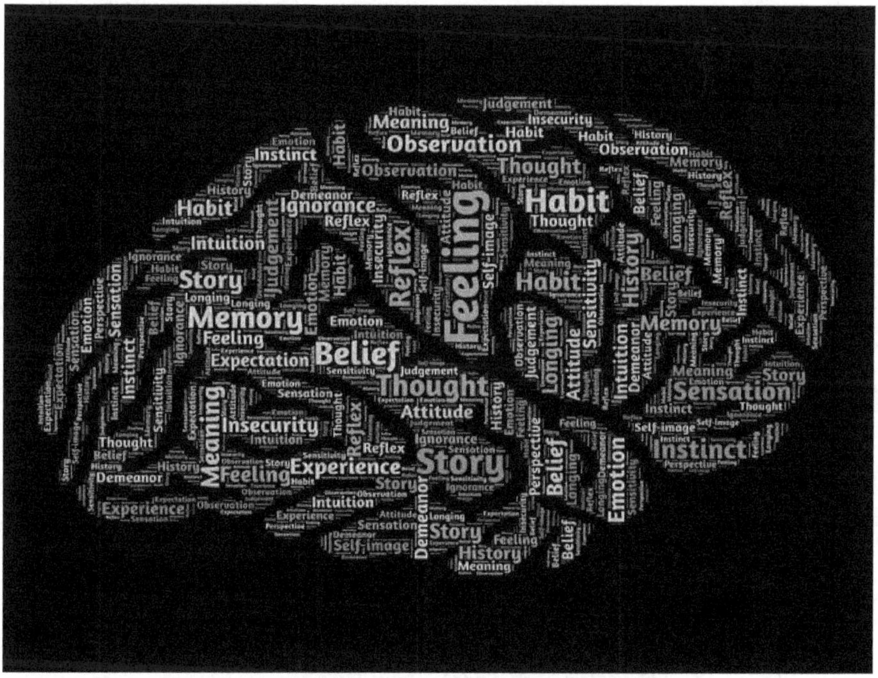

Being able to manage your mental and emotional states can be very important for your growth and your ability to overcome your social anxiety. There are a number of different methods that you can use to be able to manage these mental states as effectively as possible, which is what this chapter will be about. Several of these strategies will be detailed here.

One very popular strategy that a lot of people use to help them to manage their emotional and mental states more effectively is a very simple one; You can try putting a rubber band around your wrist, like a bracelet, in order to serve as a reminder to notice your thoughts. A lot of people will actually snap the rubber band in order to give their bodies a little bit of a shock as a punishment for negative or otherwise harmful thoughts, but in this case, simply having the band on your wrist should work just fine. You might be tempted to fidget with the rubber band every so often, as well, stretching and pulling it when your hands are not busy. This will be very helpful for the purpose of the band. Whenever you look at your rubber band (a regular bracelet can also work) you should try to take note of it and remember to pay attention to your feelings and thoughts. Often, you might catch yourself fidgeting with the rubber band out of nervousness or anxiety, in which case you can take a moment to remind yourself to try to avoid unnecessary stress or worrying too much. If you are sad, irritated, or uncomfortable and you aren't particularly good at noticing these emotional states, you might find yourself fidgeting with the bracelet in a lot of these situations, out of stress or anxiety. Try to notice whether you are uncomfortable and if you are usually uncomfortable when you fidget with your rubber band, taking note of the habits that you form around this simple tool that you now have on your wrist. When you can begin to pay attention to the habits that you begin to form with this rubber band or bracelet. You can understand the emotional states that cause those habits and any fidgeting that you might do when you are uncomfortable or irritated. When you are able to catch yourself in these negative states of mind, then you can also begin to work through those and replace them with more positive mental and emotional states instead. The rubber band is a very simple and inconspicuous tool.

Once you have spent a few days with this new rubber band or bracelet on your wrist, and you have allowed yourself to begin to understand when you mess with the band or look at it every so often and note your emotional and mental states, you can then start to actually take action with regards to these emotional responses to your surroundings. This is when you might want to pull on the rubber band and give it a little snap when you notice yourself having some sort of negative thought or feeling. If you do this, you should only snap it lightly. Of course, you are not trying to hurt yourself. You only want to draw attention to it and allow yourself to be aware of the situation by lightly snapping the band. If you do not want to actually snap that band, then you can instead switch it to the other hand or perform some other unusual action with it that you will take note of. The important thing that you need here is the actual physical pattern that you use to break the "loop" of unpleasant thoughts or feelings.

After you perform that action to break the situation, you will want to think about the situation and what led you to feel uncomfortable or unhappy, or what you might have been doing when you began to feel unpleasant. You should actively take note of this situation, and if you are alone, you might even try to say out loud what has happened. You might say to yourself, "Oh, this thing made me feel this way," or "that situation leads to these kinds of thoughts or feelings." You should be paying attention to these thoughts, allowing yourself to better understand your emotional states and your emotional responses to certain external stimuli.

If you are able to understand the specific kinds of situations and external stimuli that lead to which kinds of emotional and mental responses from you, you will be able to understand your own boundaries and where you are comfortable much more effectively.

If you keep this rubber band or bracelet on your wrist every day for a longer period of time, and you begin to form habits around the use of it as a tool to help you to recognize and understand your emotional states in this way, then you can begin to use that information to develop some more positive emotional habits and being able to replace your negative emotions with some positive ones when they come up. If you practice this for multiple days, then eventually you will be able to keep these much more positive and pleasant emotions at the front of your mind instead of other harmful or negative ones.

CHAPTER 15:BECOMING AN OPTIMIST

Finding a positive state of mind can also be helpful to your growth and your ability to overcome your social anxiety. Adopting an optimistic attitude can be very helpful in accomplishing this goal. A lot of different researchers in this topic will define it slightly differently, and this definition is usually be based on the different mentalities and the different specific areas of research that they operate within. Some of the different definitions of the term "optimism" will be listed below in this chapter for you to better be able to understand them.

One of these different kinds of definitions that people use to explain the concept of optimism is referred to as dispositional optimism. This kind of optimism is usually used to explain the degree to which a person will usually expect a lot of events that will occur in the future, or at least a majority of those events, to turn out to be positive or negative in nature. For example, if they encounter an unfamiliar object, will that person believe that the object will be inherently helpful to them or more harmful to them. Of course, a person who leans more toward the optimistic or positive end of that scale will usually anticipate more positive outcomes, where a person who tends more toward the pessimistic side of the scale will be much more negative and will perceive unfamiliar things as threats or anticipate that future events will be much more likely to bring them hard than good. Most of the time, a person who has a more positive or optimistic kind of attitude or personality will also have a much lower level of risk for things like the development of mental disorders, because they will be able to handle a lot of different kinds of situations much more effectively and will be able to maintain a much more happy and healthy kind of mental state. It can also be very commonly connected to much more healthy behaviors before and during stressful or traumatic situations as well, whereas a much more pessimistic personality will lead to or even cause much more unhealthy behaviors that end up causing a significant portion of the negative results or consequences that a pessimistic person will typically anticipate.

As is the case with a lot of the traits that can help to determine an individual's outlook and mental states, dispositional optimism and pessimism are also considered to be heritable traits in many cases. Optimism, as well as pessimism, are both very heavily impacted by factors within your environment, such as family or friends who you grow up with or the people around you. A lot of people will also believe that an optimistic attitude can be inherited in an indirect way because of influence from other kinds of heritable traits like alcoholism, personality type, or intelligence. A lot of studies conducted on this topic have shown that it can also be directly inherited from parents or family members in about one-fourth of cases of people with a very optimistic kind of personality and that it is a stable personality dimension and a predictor of life outcomes. Things like subtle chemical shifts and biological factors can impact the kind of outlook that a person will carry with regards to a pessimistic, or an optimistic personality can impact these traits as well as environmental factors that will impact the outlook that a person will have during their life. Most people will usually argue that these traits are based mostly on these kinds of environmental factors, but can be influenced by genetic factors as well, to a slightly smaller degree. A lot of people will develop changes in their outlooks during the course of their life, possibly flipping from an optimistic outlook to a pessimistic one or vice versa. This can increase the risk of a person developing mental conditions such as depression or anxiety later in life. Optimism and pessimism are often considered to be learnable traits, which can be helpful if you need to get

yourself out of a more pessimistic state of mind and become a more positive and optimistic person.

There is also a test that is used to measure where a person might fall on the scale of optimism and pessimism, which is referred to as the life orientation test, or the LOT. This test was designed to help people to be able to assess and evaluate dispositional optimism and pessimism in 1985 and has since become one of the most popular tests that is used to measure the prevalence of traits. This test is usually made up of eight different sections, with four sections that are made up of filler. The sections of the life orientation test are also split up into positive and negative ones as well, with four of them being more positive, and four of them being more negative. These different splits of the test are meant to create a more significant sense of diversity and a more accurate result from it. There are two primary revisions that have been made to the life orientation test, which are referred to as the extended LOT and the revised LOT. The revised LOT is made up of six sections instead of the original eight, which are scored on a more specific scale with five points which range from "strongly agree" to "strongly disagree," with the middle point being a neutral response. There are also four sections which are made up of filler as well, and each of the coded sections is also phrased similarly to the original, with half of the coded sections being phrased in a positive manner, and the other half being phrased much more negatively for the same reason as with the original test. The LOT-R is intended to produce slightly more consistent results than the original LOT test.

There is also another kind of test that is commonly used to measure a person and find out where they fall between optimistic and pessimistic outlooks, which is referred to as the attributional style questionnaire, or the ASQ. The ASQ will have the subject read off a list of six events, each of them coded in a specific way, with three of the six events being phrased in a positive way and three of them being phrased much more negatively. After each of the situations has been read off, the subject will be asked to come up with a possible cause for that specific event. These explanations will also be measured on a number of different scales such as whether the explanation of the event in question is external or internal, if it is changeable or stable, or if it is local to the event or a more global one. The ASQ has been modified a number of times, and different versions of the test are used by different people in different areas and have been changed out at various points in time, but most of the variations of this test are relatively similar, serving the same purpose and having the same general structure.

Optimism has been found to be connected to a person's well being, as well as pessimism, in positive and negative ways, respectively. People with optimistic personalities have been found to be a little bit less likely to develop different kinds of health conditions, varying from about 5 to 10 percent in the difference in likelihood of developing a number of different kinds of health conditions ranging from physical diseases and conditions like cardiovascular diseases to even mental ones like depression or anxiety. On the other end of the spectrum, people who have a more pessimistic kind of mentality will be a little bit more likely to develop these conditions when compared to positive or optimistic people.

Optimism and pessimism have been found to be connected to allow a person to be able to develop much more effective and healthy methods of handling and coping with different kinds of negative symptoms like pain or negative emotional states caused by specific events. An optimistic person will usually be much more likely to handle their problems and issues in a more direct way, taking more active steps and solving their problems, whereas a pessimistic person will be much more likely to give up or to avoid their problems. This response is largely influenced by the higher levels of cortisol that a pessimistic person will produce, which leads to higher levels of stress in your daily life. Pessimistic people might also have a lot more trouble with the regulation of these reactions, causing their cortisol receptors to be overloaded, leading to a lot of issues and a higher level of stress in general. Optimism can also improve the rate at which a person is able to recover from physical or mental stress, as well, because of the different kinds of ways that they will use to handle these kinds of events. However, the most significant and obvious thing that will come as a result of a more optimistic kind of personality will be much more generally happy in their day to day life.

There is another definition of the term optimism which is slightly different from dispositional optimism, called philosophical optimism. Where dispositional optimism focuses on a belief that things will usually work out for the best and the disposition that a person has, the philosophical view of optimism is more primarily focused on the perception that the present moment is the "most optimal" or the "best" state that can be, and that things are working out in the best way that they can in the present moment as well as future moments that will come. This philosophical optimism is very commonly referred to as "optimalism", and focuses on a belief that all of the events that occur at any given moment are governed by the laws of "optimization", and can also be thought of in terms of different "timelines", and that the one that we are currently on is the

most optimal" timeline, and that all of the events that occur within it are and will be the best combinations of those events that can possibly be. This can be thought of and explained with the common phrase "everything happens for a reason," which is typically used in a religious context, but can also be applicable with regards to philosophical optimism as well. This will usually be countered by pessimistic views related to realism that people with pessimistic personalities will have.

Philosophical optimalism is usually defined with this "everything happens for a reason" explanation, or with the timeline explanation that is similar to the many-worlds interpretation's theory in that there are "different timelines" that exist side by side and that our timeline is the most optimal or best version of the different combinations of events that occur and can occur in our timeline. Of course, this kind of outlook can be compatible with religion as well, by attributing the path of the timeline that we are on to choices that have been made by a god or gods, but it does not require one to stand up on its own. This kind of philosophy is simply a way that can be used to explain the natural laws of the world around us.

Psychological optimalism is representative of an acceptance of failure that you might experience and can be explained by connecting its opposite philosophical standpoint of pessimism to perfectionism, which can hold you back in a lot of cases. Perfectionism is often explained as a consistent and almost compulsive drive to try to reach unattainable goals of perfection which will almost always lead to a sense of failure and defeat. Perfectionism is usually driven by a drive to reach these goals and focusing on the overall accomplishments that you might be striving to reach, or some other kind of unrealistic goal. A perfectionist will usually ignore the limitations of human ability and still struggle to achieve perfection regardless of the reality of the situation. Optimalism can allow for a lot more failure in pursuit of a specific goal and expects that while the trend of activity is towards the positive, it is not necessary to always succeed while striving towards goals. This basis, in reality, prevents the optimist from being overwhelmed in the face of failure.

CHAPTER 16: THE POWER OF VISUALIZATION

It should be no surprise by this point, but social anxiety can be very difficult to live with and to cope with. Visualization can be a very helpful tool to allow you to be able top cope with and handle your anxiety and maybe even alleviate some of the symptoms that come with it much more effectively as well. If you are suffering from social anxiety disorder, then your ability to enjoy the things that you encounter can be severely impacted, and it can become much more difficult for you to be able to enjoy yourself in your day to day life, including things that you wouldn't even notice if you didn't suffer from this very difficult disorder and that you might even have taken for granted without it. Social anxiety can cause you to miss out on a lot of things and have a lot of regrets.

A lot of the biggest parts of finding good opportunities in your life can come from networking and knowing people who can open up certain doors for you or who can suggest certain paths. Of course, if you suffer from social anxiety, you will have more trouble with the networking and building new bridges or forming connections with new people, which can have negative effects on your career, especially if you choose a path that requires a lot of networking, as well. Luckily, there are a few methods that you can use to help yourself in building these kinds of connections and forming new professional relationships with people, to allow yourself to gain higher levels of confidence in your professional life as well as all other aspects of your life too. One of the most effective methods that you can use in order to do this is by using a technique referred to as visualization. If you suffer from social anxiety, you will probably be familiar with this concept, whether that is by name or because social anxiety will often involve a high level of negative visualization in anticipation of future events. However, the kind of visualization that you will need to be doing here is much more positive and will be used to help you to to become much more confident and improve your self-esteem. Instead of imagining yourself performing negatively or having a lot of negative things happening to you, you will instead need to begin to imagine (or visualize) positive things coming your way in order to begin to replace those negative expectations with positive ones, which can improve your self-esteem significantly.

This kind of visualization can actually be very easy to do and can be incredibly helpful to improve your self-esteem. It can also be used in a lot of different ways, for a number of different things including managing emotional pain. A lot of doctors will even use things like television or even virtual reality to help people to manage physical pain because those are both simple, light activities that are very easy to immerse yourself in and to use as a distraction from the real world. Visualization is very similar, only it is performed in your own mind as opposed to with the use of an external tool to help you with this visualization.

This kind of visualization will also usually take two slightly different forms. The first of these two forms is to imagine yourself in a different place from where you are actually at. Usually, you might be doing something you enjoy, or something that gives you more pleasure than what you are actually doing. This is sometimes referred to as the "happy place" technique and involves imagining yourself going to your "happy place," visualizing yourself in a pleasant situation, especially if you are actually in a more unpleasant or less enjoyable one. This kind of visualization can be very helpful for preventing yourself from feeling very negative emotions or thoughts, and for getting rid of high levels of anxiety by picturing yourself in a situation where these things are not a concern.

This method of visualization can be very effective for lessening the effects of symptoms that you are experiencing in the present moment, alleviating these symptoms much more effectively than you would otherwise be able to. This kind of visualization will have you try to imagine yourself in a different place from where you are, mentally removing yourself from the situation without necessarily needing to physically leave. You might try closing your eyes for a moment or simply stopping to think about your "happy place." After a few moments, you will very likely notice your symptoms becoming a little bit lighter, and before son, you will be much calmer and you can continue through the interaction, probably feeling refreshed. Doing this will help reduce your anxiety and will stop the snowball effect where your fear would just keep increasing.

The second method of visualization that you can use is to imagine events that you will be picturing soon in the near feature. This technique will usually involve stopping before an interview and mentally preparing for that by imagining it going very well, which can help you to improve your self-esteem and go into that interview more confidently than you otherwise would. You can be vague or specific, and it can work differently from person to person. This can be similar to maybe having conversations with yourself while you are in the shower, practicing and anticipating how you will react to certain situations. Some people, especially if they are on TV a lot, or they perform a lot of interviews or other kinds of similar "performances", where they might need to be able to be a little bit more socially adept than normally might stand in front of the mirror and practice responding to questions that they might be anticipating to themselves. This can be very effective for preparation and for improving your level of self-esteem and confidence before you enter the situation or situations that you are preparing for.

This method is slightly different from the "happy place" method, in that it is a very preemptive strategy that you can employ and will allow you to prevent the symptoms of your anxiety before you enter the situation, and can sometimes be much more effective for anxiety disorders such as social anxiety than the happy place strategy. This will also probably be able to come much more naturally to you if you suffer from social anxiety because you will probably be much more used to imagining these events in a negative way, and in order to do this, you will only need to do the same thing in a more positive context. You might even have no trouble at all flipping that switch. You might also, however, have a little bit of trouble switching from negative to positive visualization techniques. Regardless, once you pick it up and begin to become better at using this technique, you will find yourself becoming a little bit more comfortable in social situations as you become more adept at visualizing future situations in positive ways and conditioning yourself to become a more naturally confident person though these techniques.

You might even eventually find yourself reaching a point where you are able to look forward to social events because you will imagine much more pleasant situations and conversations with people at those events. If you can picture an interview going very well, then you will naturally be much more confident, and you will be much more comfortable when you enter those situations as well. When you are able to visualize future events in this way, then you will be able to give yourself a much higher chance of being confident when you are actually in these situations, and this can allow you to progress in your growth much more quickly, and you will very soon begin to be able to move beyond your social anxiety as well.

When you are trying to visualize these kinds of situations, you should try to start in a very calm mood, as relaxed as you can be. You should begin in a very calm state. If you need to, you should try to do some breathing exercises, because a lot of people will usually struggle with breathing difficulties when they become anxious or have panic attacks, because of the tightening of their chests and the airways. You might try to close your eyes and focus on the rate of your breathing, practicing some of the techniques that have been mentioned earlier in this book. You might try to notice particular sounds or other sensations that you are able to experience, taking in all of the different external stimuli around you. You should try to breathe slowly and pay attention to the way that you are breathing, taking more time on the exhale than on the inhale, breathing a little bit more slowly on both when compared to the way that you breathe normally. You should also try to breathe from your diaphragm, instead of from your chest, as was mentioned before as well. Continue to practice breathing in this way, and visualize the event that you are about to enter while making sure to be as specific as you can.

You should try to account for all of the different senses, like the smells, sounds, and sights, as well as the physical sensations that you might experience. Make sure to imagine positive things more than anything else. This is one the most important aspects of this kind of visualization. Be very specific and be very positive. You should try to start from the beginning, as well, picturing all of the things that you are participating in chronological order. For example, if you are out buying clothes, you might try to visualize yourself in your new outfit and thinking about how good you will look in that new outfit. You should try to exaggerate a little bit maybe even going a little bit overboard and picturing this hypothetical situation very vividly and being maybe more positive than the reality can be, even in a best-case scenario. You should try to indulge this exaggerated positivity as much as you can allow yourself to be, even if it seems a little bit ridiculous at first. This kind of excessive positivity can be very helpful in boosting your mood, which is the whole point of this exercise!

You should try to continue to anticipate these hypothetical positive events up until the actual thing happens, and even maybe during it. If something doesn't go your way, you might try to lean on your visualizations and get yourself back up. Try to recover quickly and move onward as well as you can. Try to visualize yourself after you have finished with the situation in question and reveling in your successes. You should try to imagine all of the sensations that you will feel during this event as well, accounting for all of the physical sensations like sights, smells, and touch as well as you can. You should try to enjoy smelling and tasting the food if you are preparing for dinner (this might also help you to decide on what you might like if you are the kind of person who struggles with deciding on menu items at restaurants!).

If you know that someone at a party or other kind of social gathering might like to wear a lot of perfume or cologne, imagine that as well, or the kind of vibe that you might feel during this event, remembering to imagine those things in a positive context. This will serve two purposes. The first is possibly boosting your self-esteem, and the second thing that this can do for you is that you might begin to view these things like an excessive amount of perfume that a person wears in positive contexts similar to the way that you imagined during your visualization. If you repeat this over and over, and as you continue to do this for everyday things, especially new or unfamiliar experiences, then you will begin to be able to do this naturally and form a habit out of it. It will eventually become natural for you and will allow you to more effectively and easily remove any doubts or fears you may have by reinforcing the positive outcome. Try to imagine this situation as many times as you can or as frequently as you can, but in a much more positive context than you might initially be anticipating before the actual event happens. Make sure to visualize the event one last time right before you go in and actually experience the event, and make sure that you are continuing to maintain a positive attitude because this is often when these kinds of things can begin to fall apart. You will need to practice the primary way where you can mentally remove yourself from an anxious situation. Repeat this time after time until you are able to visualize yourself somewhere else that you consider safe and free of stress or discomfort. Of course, this can take a lot of time and practice in order to master effectively, but once you do master this you will be able to go in, and your self-esteem will be passively increased over time, almost automatically.

CHAPTER 17: VISUALIZATION TECHNIQUES

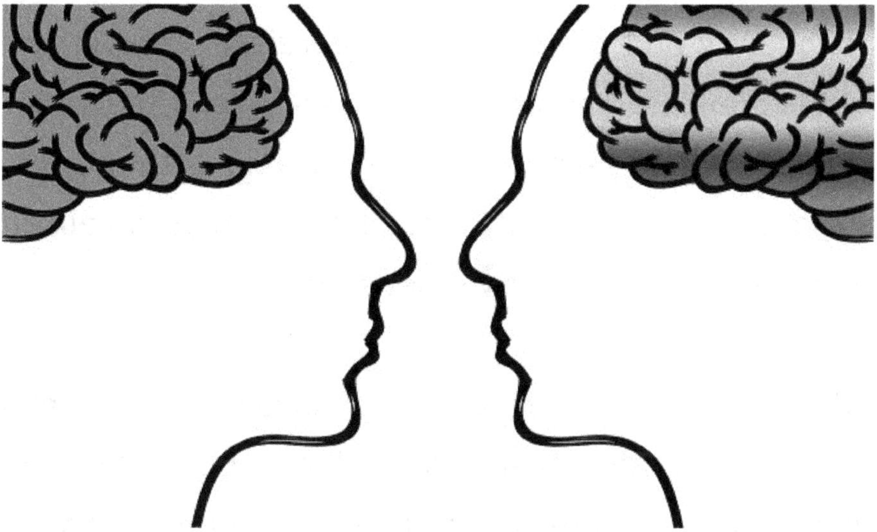

The visualization techniques that are explained in this book are very similar to the law of attraction in the way that they work and the ways that you are able to use them to improve your confidence and allow yourself to find more opportunities. The basic concept behind both of these ideas is that you will change your way of thinking by imagining positive situations that are relevant to something that you might normally picture going poorly. Replacing this kind of negativity with positivity will allow you to have a much easier time, as well a preparing you for things that are coming up in the future. If you imagine things going well, you will be preparing yourself to go into that situation much more confidently and much more comfortably. This is one of the simplest and one of the most useful tools that you can use to help yourself to overcome things like social anxiety or other kinds of stress that is caused by mental disorders or worsened by them.

When you are able to passively visualize these kinds of situations more easily, you will also be much more likely to be able to confidently handle these situations in reality. You will also be more comfortable in unfamiliar situations, being able to anticipate them much more positively, and you will be able to recondition yourself to be able to have a much more positive attitude than you would otherwise be able to.

There are a few things that you should make sure to

do as well, when you are visualizing events coming up in the future, in order to begin to excite yourself for these events. The first of these methods is to try to "fall in love with" the things that you want and the situations that you are anticipating. If you can try to artificially excite yourself about something, then that interest will eventually become real. This is similar to things like smiling to try to make yourself a little bit happier, or like taking an ironic interest or forming a real habit out of an ironic gesture or a comedic bit that just becomes a thing that you do out of habit. Your emotions and your thoughts are a very important part of your interests and desires, and they will be one of the biggest parts of being able to "reprogram" your mentality to be more positive and optimistic. However, you should also remember to be open to new possibilities and avoid becoming too attached to a particular course of events.

You should also try to visualize this fantastical situation as often as you can, but not necessarily as much as you can. This sentence might seem to contradict itself, but it essentially means that you should try to practice this technique and perform this kind of visualization in a lot of short, small bursts. If you are able to perform this kind of visualization in small sections, then you will be able to train yourself to become a little bit better at this kind of visualization, as well as being able to prevent yourself from possibly dwelling on one specific moment or getting stuck on one kind of interaction that you might be anticipating. If you can picture small, discrete moments in this way then you will also be able to anticipate a lot of different kind of scenarios with a lot more variation, and you will also eventually be able to perform this visualization on the fly, as well.

One of the most important parts of this kind of visualization is to explore different possibilities as much as you can and to try to have fun with your visualizations by exaggerating the positive imagery situations that might occur. The point of this is to form positive associations with social interactions that you are anticipating in order to build up your own confidence. You should try to imagine the best case scenario, or even a ridiculously good outcome, as long as it is a positive situation that you can enjoy and immerse yourself in. It doesn't necessarily need to be 100 percent based in reality, it just needs to be positive and allow you to enter the situation comfortably and confidently.

CHAPTER 18: MEDITATION AND BREATHING EXERCISES

Imagining your breaths carrying negative and upsetting thoughts and energy away from you or picturing waves of calmness wash through you as you exhale can also be very helpful. If you get distracted, gently bring your attention back to your breath and your words.

CONCLUSION

Congratulations! By this point you should have finished all of the different chapters in this book, and you should have a complete understanding of social anxiety disorder and the different forms that it can take, as well as the symptoms that it can cause and how you can prepare yourself from and prevent these kinds of symptoms or even how to handle them on the fly. After you have finished reading through all of the chapters in this book, you should be able to understand the symptoms and the different signs that come with social anxiety or that you might be experiencing a high level of anxiety or a panic attack. When you begin to notice these signs and symptoms, you should also be able to remember the different steps that you need to take and have a plan for how to handle your social anxiety in an effective way that you have practiced and can perform almost instinctively and without thinking. You should be able to remember the different techniques and tools that have been explained within this book as well as how to use them to help you with the symptoms that are caused y your anxiety and how to determine the undetrlying causes of these symptoms in oder to be able to handle them on a more long term basis as well as how to alleviate these symptoms for more immediate relief so that you can continue on and be able to lead a much more successful social life even if you might be suffering from a more severe case of social anxiety. With the use of these kinds of tools, you will be able to continue to maintain a normal social life and you will be able to continue to work toward moving forward and overcoming your social anxiety disorder, eventually becoming a much more mentally and emotionally stable and comfortable person, with a little bit of work.

CPSIA information can be obtained
at www.ICGtesting.com
Printed in the USA
LVHW011118301020
670159LV00014B/540

9 781914 061066